Advanced Audio Visualization Using ThMAD

Creating Amazing Graphics with Open Source Software

Peter Späth

Apress®

Advanced Audio Visualization Using ThMAD: Creating Amazing Graphics with Open Source Software

Peter Späth
Leipzig, Germany

ISBN-13 (pbk): 978-1-4842-3503-4   ISBN-13 (electronic): 978-1-4842-3504-1
https://doi.org/10.1007/978-1-4842-3504-1

Library of Congress Control Number: 2018938391

Managing Director, Apress Media LLC: Welmoed Spahr
Acquisitions Editor: Natalie Pao
Development Editor: James Markham
Coordinating Editor: Jessica Vakili

Cover designed by eStudioCalamar

Cover image designed by Freepik (www.freepik.com)

Distributed to the book trade worldwide by Springer Science+Business Media New York, 233 Spring Street, 6th Floor, New York, NY 10013. Phone 1-800-SPRINGER, fax (201) 348-4505, e-mail orders-ny@springer-sbm.com, or visit www.springeronline.com. Apress Media, LLC is a California LLC and the sole member (owner) is Springer Science + Business Media Finance Inc (SSBM Finance Inc). SSBM Finance Inc is a **Delaware** corporation.

For information on translations, please e-mail rights@apress.com, or visit www.apress.com/rights-permissions.

Apress titles may be purchased in bulk for academic, corporate, or promotional use. eBook versions and licenses are also available for most titles. For more information, reference our Print and eBook Bulk Sales web page at www.apress.com/bulk-sales.

Any source code or other supplementary material referenced by the author in this book is available to readers on GitHub via the book's product page, located at www.apress.com/978-1-4842-3503-4. For more detailed information, please visit www.apress.com/source-code.

Printed on acid-free paper

Table of Contents

About the Author

Peter Späth holds three degrees in physics: a master's from University of Texas at Austin (1994), a diplom from Universität Würzbug (1996), and a PhD from the Technische Universität Chemnitz (2000). He became an IT consultant in 2002. In 2016, he decided to concentrate on writing books on various topics, with the main focus on software development. Throughout his career he has always preferred to work with open source tools and favors operating system–independent platforms like Java. After using various audio visualization programs, he designed his own visualization suite, ThMAD, to overcome the deficiencies of important functionalities, quality documentation, and bug fixes of existing visualization suites.

About the Technical Reviewer

 **Massimo Nardone** has more than 23 years of experience in security, web/mobile development, cloud computing, and IT architecture. His true IT passions are security and Android.

He currently works as the chief information security officer (CISO) for Cargotec Oyj and is a member of the ISACA Finland Chapter board. Over his long career, he has held many positions including project manager, software engineer, research engineer, chief security architect, information security manager, PCI/SCADA auditor, and senior lead IT security/cloud/SCADA architect. In addition, he has been a visiting lecturer and supervisor for exercises at the Networking Laboratory of the Helsinki University of Technology (Aalto University).

Massimo has a master's degree in computing science from the University of Salerno, and he holds four international patents (related to PKI, SIP, SAML, and proxies). Besides working on this book, Massimo has reviewed more than 40 IT books for different publishing companies and is the coauthor of *Pro Android Games* (Apress, 2015).

Introduction

This book is a sequel to *Audio Visualization Using ThMAD* and introduces advanced examples and features not covered in that book. It is *not* mandatory to have read that book if you already have considerable insight into the ThMAD program; if you are a beginner, then you will want to pick up a copy of *Audio Visualization Using ThMAD* because it contains a complete reference of the user interface and describes all the modules available in ThMAD.

For this book, the code is based on ThMAD version 1.1.0, and the associated OpenGL version as of the time of writing is 3.0.

Targeted Audience

This book is for artists with some IT background and for developers with artistic inclinations. Also, it is for readers of the first book, *Audio Visualization Using ThMAD*, who want to improve their proficiency in using ThMAD.

Installation

After you download ThMAD from https://sourceforge.net/projects/thmad/ as a Debian package with the suffix .deb, you need to make sure the dependencies are fulfilled. A future version might do this for you automatically, but for now you have to do it manually. ThMAD depends on the following packages; entries marked with an asterisk (*) are probably already installed on your Ubuntu Linux system:

- libglfw3 ($\geq$ 3.1)

- *libc6 ($\geq$ 2.17)

- *libfreetype6 ($\geq$ 2.2.1)

- *libgcc1 ($\geq$ 1:4.1.1)

- *libgl1-mesa-glx ($\geq$ 11.2.0) or libgl1

- *libglew1.13 ($\geq$ 1.13.0)

- *libglu1-mesa ($\geq$ 9.0.0) or libglu1

- *libjpeg8 ($\geq$ 8c)

- *libpng12-0 ($\geq$ 1.2.13-4)

- *libpulse0 ($\geq$ 0.99.1)

- *libstdc++6 ($\geq$ 5.2)

To install these packages, log in as root in a terminal and enter this as one line:

```
apt-get install libc6 libfreetype6 libgcc1 libgl1-mesa-glx libglew1.13 libglu1-mesa libjpeg8 libpng12-0 libpulse0 libstdc++6 libglfw3
```

If you downloaded ThMAD via a browser to the folder Downloads in your home directory, you will install it, still as root, via the following:

```
dpkg -i /home/[YOUR_USERNAME]/Downloads/thmad_1.0.0_amd64.deb
```

Make sure this is entered as one line and replace [YOUR_USERNAME] with your Linux username. If there is a newer version available, you can give it a try, but this book is for version 1.0.0.

All the files will end up in /opt/thmad. After that, please log off as root by pressing Ctrl+D. This is important so that subsequent actions do not mess with your system.

For your convenience, launchers are available; you can find them on your desktop after the following commands:

```
cp /opt/thmad/share/applications/thmad-artiste*.desktop ~/Desktop
cp /opt/thmad/share/applications/thmad-player*.desktop ~/Desktop
```

(Enter both in one line each.) While the installation main folder can be renamed, the launchers depend on the installation residing in /opt/thmad. You could, however, edit the launchers appropriately, if you think a different installation folder is a better option for you. To see whether everything works, use the launcher thmadartiste.desktop, or in the terminal enter /opt/thmad/thmad_artiste.

As another preparation step before actually using ThMAD Artiste, you might want to consider releasing the Alt key from the operating system. The default Ubuntu window manager, Unity, uses the Alt key to start the heads-up display (HUD), but ThMAD uses it for various GUI actions. To disable Ubuntu using the Alt key for the HUD or to change the key binding, go to the Keyboard section of Preferences, advance to the Shortcuts tab, and then go to the Launchers menu. Select the entry "Key to show the HUD" and press Backspace to disable it, or choose a new key or key combination to change the binding.

Upgrading ThMAD

If you have ThMAD version 1.0 running on your system, you can easily make all your files available to the newer version 1.1. All you have to do is copy your files from /home/[USER]/thmad/1.0.0 to /home/[USER]/thmad/1.1.0.

Conventions Used in This Book

Working with ThMAD is extensively coupled with using its modules, which are organized in a treelike structure. I'll usually refer to modules like `maths` ➤ `converters` ➤ `4float_to_float4`. If the module position inside the module tree is clear from the context, I'll shorten this to `4float_to_float4`.

State is the common notion for a rendering pipeline while constructing it. Finished *states* are also called *visuals*. References to sample states, including associated code provided with the installation and informational hints in general, are highlighted as follows:

Note   This is a note. It might point you to a source file called `A-3.2.1_Visualization_basics_basic_samples_basic_2d_sample` in the `TheArtOfAudioVisualization` folder. By *folder* in this context I mean a folder showing up in the module lister or browser.

Tip   You'll also find helpful tips like this.

Caution   Important notes and pitfalls are marked like this.

Small code snippets appear directly in the text in monospaced font. Code and script snippets, as well as terminal input and output, usually show up as blocks in monospaced font like here:

```
apt-get install libc6 \libfreetype6 libgcc1 \libpulse0
libstdc++6 libglfw3
```

If a longer line does not fit onto the book page's width, a trailing ⌐ at the end of each line of code signifies that the ⌐ must be removed and the subsequent line break discarded. For example, the following:

```
echo "cmd [...] rectangle ⌐
abc [...]"
```

should be entered as follows:

```
echo "cmd [...] rectangle abc [...]"
```

In many places, an asterisk (*) is used as a wildcard to denote any string. This frequently is used to refer to all the files inside a folder or to file name patterns.

Upon the first startup, ThMAD Artiste creates a data folder for all your states and visualizations at /home/[USER]/.local/share/thmad and creates a symbolic link at /home/[USER]/thmad pointing to the states and visualizations. If I'm referring to the data folder in this book, I will provide the link location.

How to Read This Book

This book should be read sequentially. To start, Chapter 1 serves as a concise operating manual for the different parts of the suite you are going to use, and Chapter 2 delves into meshes and particle systems in more depth.

Chapter 3 handles the way advanced timing issues can be addressed in ThMAD, allowing for a kind of video workstation view of ThMAD. Chapter 4 introduces how to use shaders for visualization purposes, as well as the way ThMAD handles them. Chapters 5 and 6 contain a collection of independent tutorials that you can work through in any order, with Chapter 6 concentrating on shader constructs. Chapters 7 and 8 talk about incorporating ThMAD in a JACK or ALSA sound server setup. ThMAD can be controlled from the outside, allowing you to bypass the Artiste user interface, which is described in Chapter 9. Chapter 10 explains advanced configuration issues.

CHAPTER 1

Program Operation

This chapter describes how the two main programs of the ThMAD software suite, Artiste and Player, can be started and stopped. It also describes all the possible options for controlling the programs.

ThMAD Artiste Operation

ThMAD Artiste is the program for creating visualization sketches, called *states*, and viewing them in preview mode.

Starting and Using Different Modes

If you followed the installation instructions in the book's introduction, you will find two launchers on your desktop.

 Start Artiste in windowed mode

 Start Artiste in full-screen mode

Clicking the buttons takes you to /opt/thmad/thmad_artiste (for windowed mode) and /opt/thmad/thmad_artiste -f (for full-screen mode).

© Peter Späth 2018
P. Späth, *Advanced Audio Visualization Using ThMAD*,
https://doi.org/10.1007/978-1-4842-3504-1_1

If you started ThMAD Artiste from inside a terminal, the full set of program options is available. For details about all the options, see Table 1-1.

Here is how you set individual options:

```
/opt/thmad/thmad_artiste [option1 option2 ...]
```

Table 1-1. *Artiste Program Options*

Option	Description
<none>	Starts Artiste in windowed mode. This option shows the canvas for creating sketches, a small preview window, and a module list.
-help	Shows help and immediately quits the program.
-h	Same as -help.
-sm	Prints all detected monitors and immediately quits the program. You can use the output to specify a monitor number for the -m option.
-m mon	Uses the monitor number for full-screen mode. This has no effect if not used with the -f option.
-f	Starts Artiste in full-screen mode. It is not possible to switch to full-screen mode from inside the program. You can exit this mode by pressing the Esc key. This option can be used in conjunction with the -ff and -fn options.
-ff	Starts Artiste in full-window mode. The graphics output will use the complete window space. You can later switch back to the non-full-window standard mode by pressing Ctrl+F.

(continued)

Table 1-1. (*continued*)

Option	Description
-fn	If in full-window mode, suppresses the info text in the header area.
-s 1024x860	Sets the window size in windowed mode. 1024x860 is just an example; see the output of the -sm option for possible values.
-p 200x100	Sets the window position; 200x100 is just an example.
-novsync	Experimental; disables using double buffering.
-gl_debug	Experimental; activates special OpenGL debugging feature.
-port 3267	Starts a TCP/IP port where commands to control ThMAD from outside may be sent to. 3267 is just an example.
-sound_type_alsa	Directly uses the ALSA API instead of PulseAudio.
-sound_type_jack	Use a JACK sound server endpoint to connect.
-snd_rtaudio_ device=5	If using ALSA or JACK, specifies the sound device to use. Sound devices get listed upon startup, but the audio_visualization_listener module must be present.

By default, full-window mode shows some status information in the header area. It can be disabled or enabled by pressing Alt+T. Or you can start Artiste with the -fn option to disable the status information from the beginning.

There is also a performance mode, which presents an overlay of the state creation canvas and the graphics output. To enable it, start in full-window mode or switch to full-window mode (Ctrl+F) and then press Alt+F. You can leave performance mode by pressing Alt+F again. Also, when in performance mode, you can toggle the visibility of the header info lines by pressing Alt+T.

To leave the program in any mode, press the Esc key, or, if available, click the Close button of the window or use the main context window.

Stopping ThMAD Artiste

You can stop ThMAD Artiste via any of the following:

- Right-click an empty spot of the canvas and select Exit. ThMAD detects if you have saved your changes; if this is not the case, it will ask you whether you really want to exit.

- Press the Esc key. ThMAD will tell you if there are unsaved changes.

You can also use a module called system → shutdown to shut down from inside a rendering pipeline. You can place it on the canvas and connect it to the screen module screen0. As soon as the module's input exceeds 1.0, the program will shut down.

Artiste Files

ThMAD Artiste will look for its files in the following folder:

/home/[USER]/.local/share/thmad/[VERSION]

and in these subfolders:

- /states: From here the states are loaded, and this is where they get saved.

- /resources: Here Artiste will look for or save resources such as data files and images.

- /visuals: This is used only when exporting finished states as visuals.

- /prods: This is used only when exporting finished states as prods.

- /faders: This is used only when saving faders.

- /macros: When macros get saved, they will go here.

During installation, a link to the folder /home/[USER]/.local/share/thmad gets created at /home/[USER]/thmad for convenience.

The single configuration file used by Artiste for defining some settings is located here:

/home/[USER]/thmad/[VERSION]/data/thmad.conf

Chapter 9 explains more about the configuration.

ThMAD Player Operation

With ThMAD Player, you can play visuals, which are exported states from Artiste. This happens when you invoke the Compile ➤ Music Visual command from Artiste's main context pop-up.

Starting and Using Different Modes

In default operation mode, ThMAD Player will recursively register all the visuals it finds inside the user's data folder and play them one by one. If you have used ThMAD's predecessor VSXu, where ThMAD Player by default looks in the installation folder, you should be aware of this difference.

5

Also, contrary to Artiste's operation, Player knows how to handle *faders,* which introduce a transition between visuals when it comes to switching from one to another. Faders are also created inside Artiste and from there exported via Compile ➤ Music Visual Fader. They then end up inside the `faders` folder.

Player will, however, not see your exports automatically since the data spaces for Player and Artiste are kept separate. To make exported states available to ThMAD Player, you have to copy the visuals and possibly the faders from the following locations:

```
/home/[USER]/thmad/ [VERSION]/data/visuals
```

```
/home/[USER]/thmad/ [VERSION]/data/faders
```

and put them in the following locations:

```
/home/[USER]/thmad/ [VERSION]/data/player_visuals
```

```
/home/[USER]/thmad/ [VERSION]/data/player_faders
```

Alternatively, you can copy them to some other place and tell Player via a startup option where to find them (using the `-path` flag).

If not using a launcher but instead the terminal to start Player, you can use all the available options, as shown in Table 1-2.

Here is how to set certain options:

```
/opt/thmad/thmad_player [option1 option2 ...]
```

Table 1-2. *Player Program Options*

Option	Description
<none>	Starts Player in windowed mode. It also recursively loads all visuals from the following path: /home/[USER]/thmad/ [VERSION]/data/ player_visuals and uses all faders found in this folder: /home/[USER]/thmad/ [VERSION]/data/ player_faders Visuals and faders are played in random order, each running for 30 seconds. Note that /home/[USER]/thmad is a symbolic link to the following location: /home/[USER]/.local/ share/thmad
-help	Shows help and immediately quits the program.
-h	Same as -help.
-path PATH	Does not load the visuals from the local user data path; see the <none> options. Instead, it loads all the visuals from the following path: PATH/player_visuals and uses all the faders found in this folder: PATH/player_faders
-dr	Disables the randomizer. Player will then not automatically cycle through the available visuals. Still, the visual chosen will be a random one.

(continued)

Table 1-2. (*continued*)

Option	Description
-rb 20	Sets the number of seconds to wait before changing to the next visual, if the randomizer is *not* disabled. If this option is not given, the value defaults to 30 seconds.
-rr 10	Randomizes the randomizer, if not disabled. A random visual duration will be chosen between the base number from the -rb option and the -rb number plus the -rr value. In this example, it's between 20 and 30 seconds. If this option is not given, the value defaults to 0 seconds.
-f	Starts in full-screen mode.
-sm	Lists available monitors and monitor modes.
-m 2	If in full-screen mode, uses monitor 2 in this example.
-fm	Lists available video modes for full-screen mode. Depends on the monitor chosen (see the -m option).
-p 300x200	If in windowed mode, sets the window position to (300;200) in this example.
-s 640x480	If in windowed mode, specifies the window size. 640×480 is only an example; choose any size you like. If in full-screen mode, this may be used to set the resolution. ThMAD then tries to find the best possible match. See the -fm option for a list of available video modes. If this is not given and the full-screen mode and possibly some monitor are requested, the video mode will automatically be chosen based on your current settings. Letting the system choose is the preferable way.
-no	Specifies no splash screen and overlay. This means Player will start immediately with the first visual, and it will not print a visual's name at its beginning.

(*continued*)

Table 1-2. (*continued*)

Option	Description
-lv .	Lists visuals seen by the Player. This depends on the -path option if chosen.
-lf	Lists faders seen by the player. This depends on the -path option if chosen.
-port 3267	Starts a TCP/IP port where commands to control ThMAD from outside may be sent to. 3267 is just an example. The details of the protocol are not part of this book.
-sound_type_ alsa	Directly uses the ALSA API instead of PulseAudio.
-sound_type_ jack	Uses a JACK sound server endpoint to connect.
-snd_rtaudio_ device=5	If using ALSA or JACK, specifies the sound device to use. Sound devices get listed upon startup, but the audio_ visualization_listener module must be present.
-schedule <S>	Specifies a schedule; see the next section for more information.

Unlike ThMAD Artiste, in ThMAD Player the visual will immediately cover the whole window or screen, and there is no context menu for the player. You can, however, press F1 to get some basic on-screen help.

Note   If you request a certain resolution in full-screen mode, it may cause the ThMAD program to terminate and show your desktop in that new resolution. You may have to manually revert the resolution setting or restart your desktop if you want to switch back to the resolution you are accustomed to.

Scheduling the Player

You can create a storyboard by using the -schedule switch, which was introduced earlier in the chapter. The syntax is as follows:

```
-schedule ind1:s1,ind2:s2,ind3:s3,...
```

Here, indN is an index in the alphabetically sorted Player files, and sN is the number of seconds as a floating-point value. The list is zero-based and the same as seen by the outcome of the -lv switch.

For example, if inside the visuals directory you have three visuals (called A_visual1, B_visual2, and C_visual3) and one storyboard that says "play B_visual2 for 15.4 seconds, then A_visual1 for 10 seconds, then C_visual3 for 102.4 seconds, and then again B_visual for 45.1 seconds," you'd write the following:

```
/opt/thmad/thmad_player -schedule 1:15.4,0:10.0,2:102.4,1:45.1
```

Here, A_visual1 has index 0, B_visual2 has index 1, and C_visual3 has index 2 in the list. You can also write it as one line, in which case you'd omit the backslash.

Stopping the Player

You can stop ThMAD Player with either of these actions:

- Press the Esc key with the focus on the ThMAD Player window. In full-screen mode, no focus is needed.

- If while constructing the state you placed the module system → shutdown on the canvas and connected it to screen0, as soon as the module's input exceeds 1.0, the program shuts down.

Player Files

ThMAD Artiste will look for its files in the following folder:

`/home/[USER]/.local/share/thmad/[VERSION]`

Alternatively, it will look in [PATH] if the option -path is specified, as mentioned earlier. It will also look in these subfolders:

- `/player_visuals`: These are visuals Player will show.

- `/player_faders`: These are faders the Player will use for transitions between visuals.

Summary

In this chapter, you learned how to invoke ThMAD Artiste and Player and what options you have when you use the starters from inside a terminal. You saw that Artiste can run inside a window or cover the whole screen. It can mix input and output on the same screen.

In the next chapter, you will learn about meshes and particle systems in detail.

CHAPTER 2

Insight into Meshes and Particle Systems

Meshes and particle systems are the two ways ThMAD introduces 3D into sketches. In this chapter, you will investigate both thoroughly.

ThMAD Meshes

A mesh in ThMAD is an advanced concept tailored for the high-performance rendering of 3D objects. It is a conceptual extension of OpenGL vertices; while OpenGL's drawing primitives consist of points, lines, triangles, and quads in various alternatives, meshes add *faces* as a concept for the precalculation of certain properties. This allows additional functionalities for the renderers. Also, in the rendering steps, meshes by default store and handle data on the graphics hardware using vertex buffer objects. Of course, OpenGL alone is capable of using VBOs, but the *meshes* in ThMAD take away some of the technical burden of defining and using them.

In detail, a mesh inside the ThMAD application consists of the following data: vertices, vertex normals, vertex colors, vertex texture coordinates, faces, face normals, and vertex tangents. I describe all of these in the following sections.

© Peter Späth 2018
P. Späth, *Advanced Audio Visualization Using ThMAD*,
https://doi.org/10.1007/978-1-4842-3504-1_2

Vertices

Vertices are arrays of three-dimensional position vectors consisting of the points that build the mesh.

These are the same vertices as in OpenGL. Vertices are the corner points defining the flat parts of a surface. This makes immediate sense if the objects consist of flat surface parts, as shown in Figure 2-1.

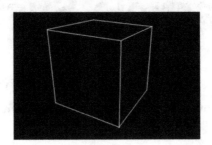

Figure 2-1. *A mesh built of flat faces*

With round shapes, you can think of them as being composed of many small, flat, surface atoms, so from a computational perspective, round shapes consist of vertices, as shown in Figure 2-2.

Figure 2-2. *A mesh built of round shapes*

Vertex Normals

Vertex normals are arrays of three-dimensional normal vectors, one for each vertex. Normals are needed for lighting. Take, for example, a surface point hit by a light beam at some angle. With the camera position looking at some other angle at that point, the graphics hardware needs to know the orientation of the surface to determine how the light beam gets reflected, as shown in Figure 2-3. Such a normal gets calculated by interpolating between adjacent vertex points.

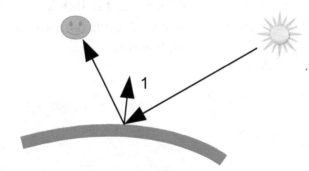

Figure 2-3. *Lighting calculation using normals*

At first sight, it might seem reasonable to tell the graphics hardware about such normals perpendicular to surface parts. Say, for example, you have a triangle ABC built from three vertices; you could add the normal vector $\vec{n}_s$ to the plane as an aide or to the graphics hardware to calculate the proper lighting. There is a problem with this approach, though. Consider another triangle ABD having the vertices A and B in common with the first one, but at an angle to it. Now, assigning its own normal vector $\vec{n}_s'$ to ABD, the lighting engine using just those two vectors for the two triangles would calculate a constant lighting L_1 for all ABC and would calculate another constant lighting L_2 for all ABD. At the edges, however, you would have an abrupt transition. This is not a problem if you have a body built of flat surface parts like a box, but with

15

a round shape it would create some ugly artifacts at the edges of the small surface parts approximating the round shape.

There is a clever remedy for this. Instead of adding normals to surface parts, the normals get added to vertices. And for a particular surface part, say again that triangle ABC, you don't necessarily use three equal normals, but instead use different normals ($\vec{n}_A$, $\vec{n}_B$, and $\vec{n}_C$) at each vertex. For each point inside ABC, you let the graphics hardware *interpolate* the normal given the three vertex normals, as shown in Figure 2-4. That way, if you do the same for adjacent surface parts like the triangle ABD, you will achieve a smooth transition of the lighting at the edges, and you will simplify the data handling by just adding one normal to one vertex, disregarding what kind of surfaces it is part of.

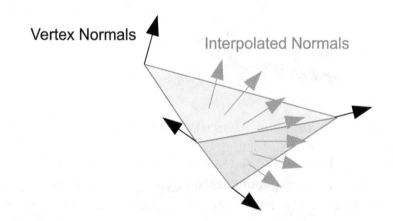

Figure 2-4. *Smoothly interpolated normals*

From an application perspective, the faces or surface parts are known, including their normals, and the vertex normals can be calculated, for example, by averaging the adjacent face normals.

Vertex Colors

Vertex colors are arrays of colors assigned to vertices, one color per vertex. A rendering process might use these values to calculate the color of each point of the surface defined by the vertices via interpolation, as shown in Figure 2-5. (For this to work, the `vertex_colors` anchor of the `mesh_basic_render` module must be set to yes). Not too rarely, however, these values are misused for various other purposes.

Figure 2-5. *A mesh face with interpolated colors*

Note that if you are using textures, you usually do not need vertex colors, although it is not forbidden to mix textures and vertex colors using some blending scheme.

Vertex Texture Coordinates

Vertex texture coordinates are arrays of texture coordinates, one assigned to each vertex. The original purpose of textures is to use a two-dimensional image and span it over the surface parts of a 3D body. For this to work, you need to know which vertex belongs to which point inside the texture, as shown in Figure 2-6.

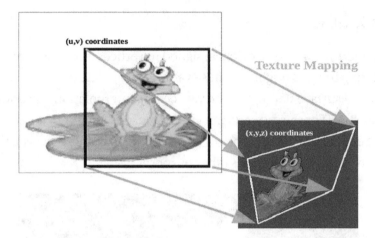

Figure 2-6. *Texture mapping*

Since a renderer might decide quite freely what to do with texture coordinates, these values are subject to being misused for other purposes as well.

Note that although the mapping to texture coordinates primarily is a mapping of 3D coordinates to two-dimensional texture position vectors, OpenGL allows for other dimensionalities.

This advanced texture mapping technique, however, plays no role in ThMAD.

Faces

Faces are precomputed atomic pieces of the surface defining the mesh. ThMAD uses faces internally; they do not directly correspond to OpenGL objects.

Face Normals

Face normals are the normals to faces. They have the same meaning as the vertex normals described earlier but are associated to the faces.

18

This feature is not used often since OpenGL depends on the vertex normals. However, a plug-in might calculate or use face normals as an intermediate step to eventually provide for the vertex normals.

Vertex Tangents

The module mesh → modifiers → helpers → mesh_compute_tangents may be used to calculate the tangents for all faces and store them in the mesh. In mathematics, a tangent is a straight line that has exactly one point in common with the convex point set that it is the tangent *for*, without crossing it.

This is not directly applicable for flat faces, but a somewhat similar concept is used: a tangent in the current context is a quaternion describing the rotation around the normal by any angle. It is not used by OpenGL, but ThMAD uses it for some mesh-related modules for special effects.

Vertex Buffer Objects

The standard mesh renderer renderers → mesh → mesh_basic_render utilizes *vertex buffer objects* (VBOs) . This is a means to store and handle vertex data directly on the graphics hardware, which gives an enormous performance boost compared to handling vertex data in the CPU's memory space.

Upon the initialization of mesh_basic_render, the following happens:

1. The vertex normals are stored in an array on the graphics hardware. This happens by calling the OpenGL function glBufferSubDataARB().

2. The texture coordinates are stored in an array on the graphics hardware, using the same OpenGL function.

3.   If per-vertex colors are available, they are stored in
     an array on the graphics hardware, using the same
     OpenGL function.

4.   The vertices themselves are stored in an array on
     the graphics hardware, using the same OpenGL
     function.

All the other data from the mesh lives in the CPU space and is not
directly shared with the graphics hardware. While the rendering takes
place, the graphics hardware just needs to be told where to find vertices,
normals, and colors, and it can thus work very efficiently without each
frame being sent all the data. The OpenGL function used for that is
glDrawElements().

A Box Mesh in ThMAD

The most basic meshes in ThMAD define vertices, vertex normals, and
faces, where the faces are just integer pointers into the vertex array.
Looking at a basic sample, a cube of size 2×2 around the origin is given by
eight vertices and six faces, as follows:

- *Front top left*: (-1, 1, 1)

- *Front top right*: (1, 1, 1)

- *Front bottom left*: (-1, -1, 1)

- *Front bottom right*: (1, -1, 1)

- *Back top left*: (-1, 1, -1)

- *Back top right*: (1, 1, -1)

- *Back bottom left*: (-1, -1, -1)

- *Back bottom right*: (1, -1, 1)

See Figure 2-7 (the vertices are numbers, and the faces are capital letters). From a mathematical point of view, you are done defining the cube. But, as mentioned earlier, you need the faces to have a proper mesh. The faces are telling what vertices they consist of, as shown here:

- *Front*: 1 - 3 - 4 - 2

- *Top*: 1 - 2 - 6 - 5

- *Right*: 2 - 4 - 8 - 6

- *Left*: 1 - 5 - 7 - 3

- *Bottom*: 3 - 7 - 8 - 4

- *Back*: 5 - 6 - 8 – 7

The order for the vertices per face is important; the order follows the *right-hand rule*. If you let your bent forefinger of your right hand follow the vertices, your thumb must show *away* from the 3D object.

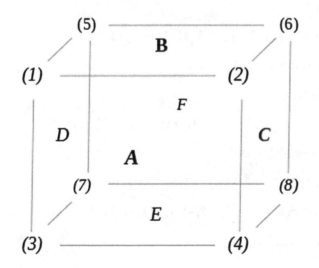

Figure 2-7. *A cube mesh*

You still need the normals per vertex, and there you encounter a problem with the cube mesh shown in Figure 2-7. Vertex (1), for example, is part of faces (A), (B), and (D), and each of them has its own normal vector pointing perpendicularly away from it. For that reason, you need each vertex three times, summing up to 24 vertices in total.

So, you clone each of the vertices twice, and eventually you end up with the following mesh definition of the cube:

- *Face A*:

    Vertices (-1, 1, 1) (-1, -1, 1) (1, -1, 1) (1, 1, 1)

    Normal: (1, 0, 0)

- *Face B*:

    Vertices (-1, 1, 1) (1, 1, 1) (1, 1, -1) (-1, 1, -1)

    Normal: (0, 1, 0)

- *Face C*:

    Vertices (1, 1, 1) (1, -1, 1) (1, -1, 1) (1, 1, -1)

    Normal: (1, 0, 0)

- *Face D*:

    Vertices (-1, 1, 1) (-1, 1, -1) (-1, -1, -1) (-1, -1, 1)

    Normal: (-1, 0, 0)

- *Face E*:

    Vertices (-1, -1, 1) (-1, -1, -1) (1, -1, 1) (1, -1, 1)

    Normal: (0, -1, 0)

- *Face F*:

    Vertices (-1, 1, -1) (1, 1, -1) (1, -1, 1) (-1, -1, -1)

    Normal: (-1, 0, 0)

Such a cube mesh with material properties and lighting might then look like Figure 2-8.

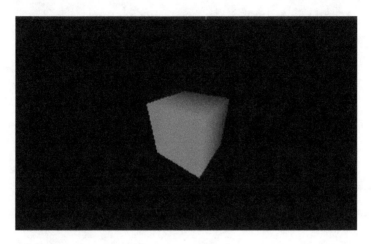

Figure 2-8. *Elaborated 3D scene*

ThMAD Particle Systems

Particle systems are about hundreds, thousands, or maybe many more small 3D objects entering a scene; obeying some generation, movement, or interaction; and finally decaying rules. You use them to simulate fog, waterfalls, candles, bursts, interesting artificial effects, and things like that.

They came to life because of greatly improved graphic hardware capabilities during the past few decades, and with the modern state-of-the-art or just medium-grade graphics and computation hardware, it is possible to have hundreds of thousands of particles at work. There is no genuine particle system concept you can find inside OpenGL; particles are just small 3D objects.

Fortunately, ThMAD provides some modules for dealing with particle systems. Internally tuning the way OpenGL gets used for drawing the particles is crucial since there are so many 3D objects to be handled. More specifically, the modules in ThMAD do the following:

- Generate particles, either given a single point or given a set of points.

- Modify particles and their trajectories. There is some randomization of particle size, precisely controlling all particle sizes, applying wind, letting them act like fluid particles (this is broken currently), letting them follow gravity, letting them bounce, and letting them rotate.

- Control a particle's life span.

- Control the way particles get rendered.

The following is a detailed description of particle system–related concepts.

Particle States

A particle internally has a state consisting of the following:

- A 3D vector describing its current position in space.

- A 3D vector describing the point where a particle came to life.

- A 3D vector describing the particle's current speed.

- A starting color assigned to the particle when it came to life.

- An ending color assigned to a particle for when it is going to die. The color during its lifetime gets interpolated between the starting color and the ending color.

- A current size.

- A size assigned to a particle when it came to life.

- The lifetime of a particle.

- A flag telling whether a particle is still moving or grounded and not moving any longer.

- A flag telling whether particles are rotating (experimental).

- If rotation is enabled, the current rotation angle.

- If rotation is enabled, an angle describing the rotation for a frame (1/60 second).

As for the movement state of particles, they may be alive and moved and drawn, or they may be dead and hidden and possibly waiting for revival.

Generating Particles

There are three modules for generating particles.

- `particles → generators → basic_spray_emitter`

- `particles → generators → bitmap_to_ particlesystem`

- `particles → generators → particles_mesh_spray`

The basic spray emitter uses a single point in space where particles get emitted, like for a spray. You can specify a number of properties using the anchors.

- The total number of particles to handle. This is the sum of particles alive and dead.

- The number of particles per second. This is the *maximum* number of particles to revive, i.e., to spray out from the emitting point per second. Particles are subject to getting sprayed out when the amount lies within this generation rate and the particle has been dead before. If you set this to a number less than 0, all particles that have died will be immediately revived and re-emitted from the spray position.

- The emitter position where particles get sprayed out. This position can be moved during the visualization, but the particles obey their independent movement laws once emitted.

- The emitting speed.

- The emitting speed type, either `directional` or `random_balanced`. For `directional`, just use the speed as given here:

```
speed' = speed
```

For `random_balanced`, for each coordinate, take a uniformly distributed random number:

```
speed'_x = rnd[-speed_x/2 ; +speed_x/2]
speed'_y = rnd[-speed_y/2 ; +speed_y/2]
speed'_z = rnd[-speed_z/2 ; +speed_z/2]
```

- A particle base size. Note that making both the particle size and the number of particles too high might use up all the CPU power! Be cautious.

- A particle size random weight. The emitted particle's size will be the base size plus a randomly chosen number from `rnd[-random_weight/2; +random_weight/2]`.

- A flag if rotation is to be enabled. If rotation is switched on, you can get some flickering effect, most noticeable if you also have a light switched on.

- If rotation is enabled, the rotation speed (more precisely, the rotation angle of each frame or 1/60 second).

- A particle lifetime base and a particle random weight. The calculated lifetime will be as follows:

```
base + rnd[-random_weight/2; +random_weight/2]
```

- A color.

Figure 2-9 shows the spray beam speed modes. Unfortunately, the module doesn't allow for a directed random spray beam like with a spray can. There are two workarounds, though, if you need that. The first is to instead take the module `particles_mesh_spray` described later in this chapter and use a simple object like a box with a size set to zero. That module does allow for a speed offset. The other workaround uses the `basic_wind_deformer` module described later in this chapter. It does exactly that: it adds some extra motion.

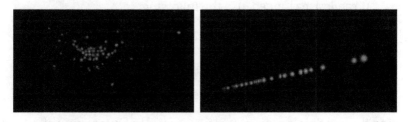

Figure 2-9. *Basic spray emitter modes*

The mesh spray emitter uses the vertices of a mesh for the positions where particles principally start their life. You can again specify the total number of particles and the revival rate. In addition, the following properties can be defined:

- The way the vertices get picked from the mesh, sequentially or randomly

- A center for spreading

- A deviation for spreading

- A spreading added to the spray position:

  ```
  actual_position = center
      + vertex_pos * spread
      + rnd[-deviation/2 ; +deviation/2]
  ```

- The speed s, a speed multiplier m, a speed randomizer r, and a speed offset f. Note that there is a bug in ThMAD version 1.0.0 for the speed multiplier formula; it's better to always set it to 1.0. As a workaround, you can always multiply the speed itself using a controller module.

- The speed type. If directional, the calculated speed is as follows:

  ```
  speed' = m * s
  ```

  If random_balanced, use the following:

  ```
  speed' = m * s * rnd[-r/2 ; +r/2] + f
  ```

  If mesh_beam, use the following:

  ```
  speed' = m * normalize[vertex_pos] + f
  ```

Using this module, you can achieve a couple of interesting effects. You can simulate point sources by using a simple mesh and setting its size to zero, linear sources like the rim of a big waterfall, round sources like disco balls, and many more. See Figure 2-10.

Figure 2-10. *Mesh spray emitter modes*

The bitmap emitter uses all the pixels of a bitmap for emitting particles. As for the other generators, you can specify the total number of particles and the revival rate. Other than that, the following properties can be defined:

- *Bitmap size*: This is not the bitmap size in pixels, but the size of the rectangle in the model space where the bitmap gets draw into.

- *Bitmap position*: This is where in space the bitmap's center gets placed.

- *Bitmap normal*: This is the normal vector of the bitmap drawing rectangle. It's part of the orientation of the bitmap.

- *Bitmap upvector*: This specifies the bitmap's rotation angle around the normal. Normally a vector is perpendicular to the normal, but you can use other vectors to add distortion effects.

- *The time source*: You can set whether to take the absolute system time or the sequencing time.

- *Other*: The speed, speed type, sizing, and rotation are the same as for the basic_spray_emitter module.

The bitmap generator allows for things such as blurring images, but you don't need to use an image bitmap; you can also use simple shapes such as circles to define particle sources. Figure 2-11 shows examples.

Figure 2-11. *Different bitmap emitters*

Modifying Particles

Particles, once emitted by a particle generator, may be subject to the modification of trajectories, size, and color.

ThMAD has the following modules for particle modification operations:

- particles → modifiers → basic_gravity

- particles → modifiers → basic_wind_deformer

- particles → modifiers → floor

- particles → modifiers → size_mult

- particles → modifiers → size_noise

- particles → fractals → ifs_modifier

While the modifier modules basic_gravity and basic_wind_deformer apply some physical force laws to the particles, the floor module imposes a motion blocker, the size_mult and size_noise modules allow for changing the flying particles' sizes, and the `ifs_modifier` module performs a one-time position jump to the coordinates of an iterated function system (IFS). You will learn more in the following sections.

Physical Law Modifiers

The `basic_wind_deformer` module applies a wind force upon the particles by adding a constant speed. This is not totally correct, because in reality particles will not immediately have the same speed as the wind but will only after some time. For small particles, though, this method gives us a fair approximation. This module is simplistic here and adds just a constant speed, doing this to whatever is given to them according to their other motion states.

The other module, `basic_gravity`, is a little more involved. You can define the center of gravity, the gravitational force for each dimension, a friction for each dimension, and a mass calculation flag. Here are some more details:

- *The gravitational center*: Since the gravitational force can be tailored independently for each dimension, you can achieve some pretty funny and unrealistic things such as a gravitational plate or a gravitational string.

- *The gravitational force as a vector, here called the amount*: For the normal physical gravitation, just set all components to the same number. If you use different numbers for the vector coordinates, each coordinate is the gravitational force contribution for the x-, y-, and z-dimensions. See Figure 2-12 for different options. The topmost shows equal contributions, the middle one zeros one, and the bottom one zeros two components.

- *A frictional component as a vector*: If not all its parts zero, this component applies a friction, slowing down the particles while they are moving along their trajectory. If you want to have realistic friction, set all components of the vector to the same number. Using different numbers will apply different frictional components for the x-, y-, and z-dimensions.

- *The time source*: This is either the real time or the sequencer time.

- *A mass calculation method*: In the real world, the acceleration of each small particle is independent of their mass. But if you want to introduce some extra effect, you can make the acceleration of each particle toward the center of gravitation depend on the particles mass, which is just the size here. Or you can give them all the same mass, which in effect does not differ from multiplying the gravitational force by space number.

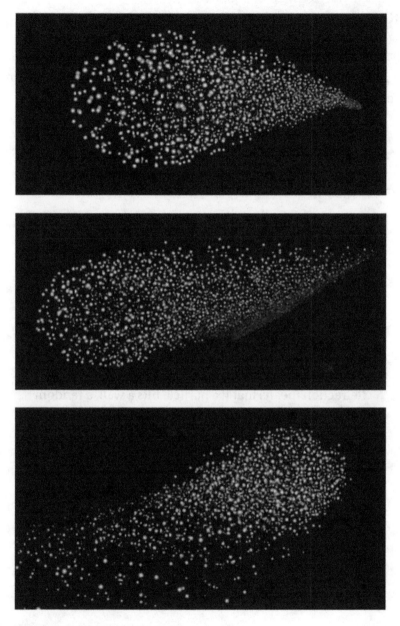

Figure 2-12. *Different gravitation modes*

Particles Hitting Walls

The `particlesystems` → `modifiers` → `floor` module introduces walls; when particles hit the walls, they either stop their movements or bounce off. Here are some details:

- An x-, y-, or z-wall or a combination thereof.

- The x-, y-, or z-position of each of the walls.

- A flag telling whether particles bounce at each wall.

- A percentage of a particle's momentum loss when hitting the walls. A loss of 100 percent means a particle does not bounce, even when bouncing is switched on. A loss of 0 percent means bounced particles will have their full speed, but in the opposite direction. The momentum loss has no meaning and is switched off when bouncing is disabled.

- A flag telling whether refraction happens at the walls. Refraction means that if a particle hits a wall, a random amount of speed gets added to the coordinates that are perpendicular to the wall's normal vector. This kind of imitates rough walls where the bounce-off angles deviates from the impact angle.

- If refraction is enabled, the refraction amount (in other words, the momentum extent of the perpendicular components).

For a couple of bouncing modes, see Figure 2-13.

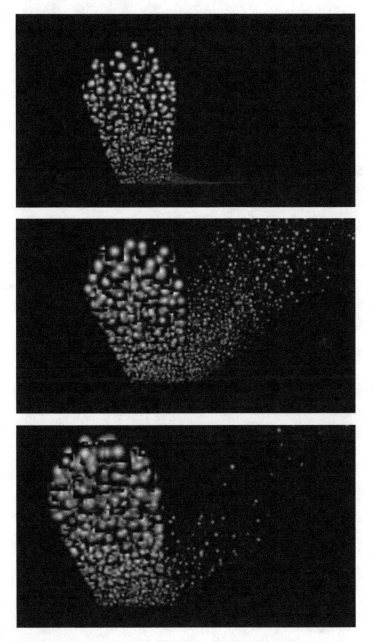

Figure 2-13. *Floor bouncing modes (top to bottom: no bouncing, normal bouncing, bouncing with refraction)*

Size Modifiers

The following two modules allow for additionally augmenting or multiplying a particle's size by some number and augmenting or multiplying the size by some random amount:

- `particles → modifiers → size_mult`
- `particles → modifiers → size_noise`

Iterated Function System Fractal

An iterated function system basically takes a point, P, and a function producing another point, $f(P) → P'$, and applies f recursively often.

$$P → f(P) → f(f(P)) → f(f(f(P))) → …$$

In this case, with $P = P(x,y,z,1)$ in homogeneous coordinates and f randomly switching between two matrices, you have the following:

$$\begin{pmatrix} a_{11} & a_{12} & a_{13} & a_{14} \\ a_{31} & a_{32} & a_{33} & a_{34} \\ a_{41} & a_{42} & a_{43} & a_{44} \end{pmatrix}$$

and the following:

$$\begin{pmatrix} b_{11} & b_{12} & b_{13} & b_{14} \\ b_{31} & b_{32} & b_{33} & b_{34} \\ b_{41} & b_{42} & b_{43} & b_{44} \end{pmatrix}$$

More precisely, you have this:

$$f: (x, y, z, q) → (a_{ij}) · (x, y, z, q) = (x', y', z', q')$$

or you have this:

$$f: (x, y, z, q) → (b_{ij}) · (x, y, z, q) = (x', y', z', q')$$

The module `particlesystems` → `fractals` → `ifs_modifier` takes the coordinates of incoming particles and applies one of two matrices on it. Specifically, its parameters are as follows:

- *A change probability*: This applies one of the matrices in each frame only occasionally. If you set this to 0.0, the IFS modifier is effectively disabled; if you set it to 1.0, all the points will be governed by the IFS all the time. You may want to set it to something like 0.1 to preserve something of the original particle system generator characteristics.

- *The two matrices*: The module will switch between the two matrices randomly, giving each an effective weight of 0.5—none of them gets preference over the other. Note that not all matrices will generate meaningful IFSs.

- *An action-selectable "change random" that produces two random matrices*: Because of the nature of IFSs, you might have to try often to get a nice IFS.

- *An action-selectable save_params*: This will save the current matrices in a file in the following folder:

  `[DATA_FOLDER]/resources/ifs`

- *An action-selectable load_params*: This will let you choose one of the previously saved IFSs. Figure 2-14 shows an example.

Figure 2-14. *Iterated function system*

Summary

In this chapter, you learned more in-depth details about ThMAD meshes and ThMAD particle systems.

The next chapter covers advanced timing options in ThMAD.

CHAPTER 3

Timing

ThMAD has two concepts of time. One is the operating system time, which is like a wall clock that is totally external and cannot be adjusted. The other is the *normal* or *sequencing* time, which is under the control of sequencing modules and will explained in this chapter. You can control this sequencing time using modules and can stop and start it. Once the time is stopped, you can even set it manually to an arbitrary value.

Engine States

So that you understand this a little bit better, I first will explain engine state. When Artiste or Player starts, the engine first finds itself in state ENGINE_ LOADING . Then it loads all the modules it sees, and after that has been accomplished successfully, it switches to state ENGINE_PLAYING. If later a trigger to stop the engine gets fired, it switches to state ENGINE_STOPPED.

In addition, a module may request a rewind, which switches the engine to state ENGINE_REWIND. This automatically yields a subsequent ENGINE_STOPPED.

Once in state ENGINE_STOPPED, a module can start the engine again by requesting a change to ENGINE_PLAYING. See Figure 3-1.

© Peter Späth 2018
P. Späth, *Advanced Audio Visualization Using ThMAD*,
https://doi.org/10.1007/978-1-4842-3504-1_3

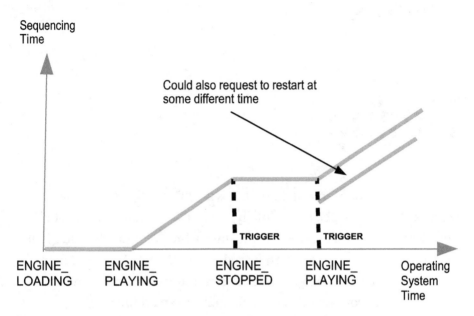

Figure 3-1. *Engine states*

Note that these states have no influence on the rendering process itself, so the engine in state ENGINE_STOPPED will still calculate the screen output.

In any of these states, the operating system time runs without any influence upon it, and if your rendering pipeline is tailored to use only the operating system time, you won't see any direct influence of these engine state changes. You'll see an indirect influence, because drawing starts only after all modules have loaded. If, however, you use the sequencing or normal time, in ENGINE_STOPPED state the sequencing time stops running and later starts running again only after an explicit switch to ENGINE_ PLAYING is performed. Also, if in stopped state, the actual sequencing time can be adjusted manually via setting trig_set_time. See Figure 3-2. A certain module can do that, which you'll learn about in a minute. The sequencing time used after a restart from a stop event (when not manually changing the trig_set_time value) is exactly the time when it was stopped. See Figure 3-3.

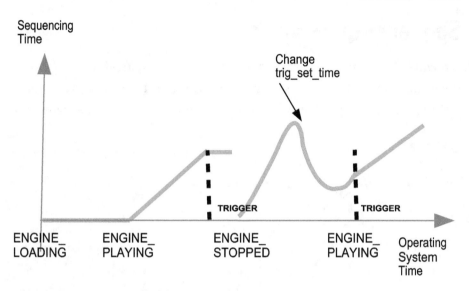

Figure 3-2. *Manually changing the sequencing time*

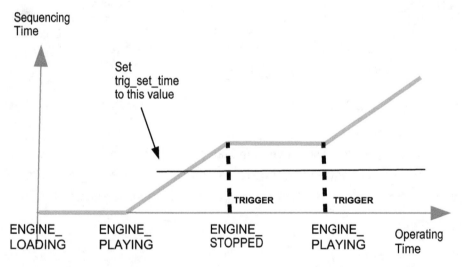

Figure 3-3. *Normal resuming*

41

Sequencing Rewinds

A rewind request will immediately switch the sequencing time to zero or
the value of `trig_set_time`, whichever is bigger, and only then stops the
engine. Resuming then happens from this sequencing time. See Figure 3-4.

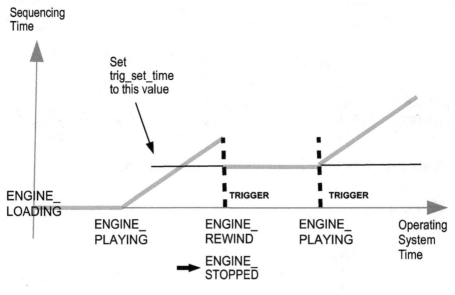

Figure 3-4. *Rewind*

If you first issued a stop event and later a rewind event, the rewind
event will set the sequencing time to zero or the value of `trig_set_time`,
whichever is bigger. Resuming then starts at this value. See Figure 3-5.

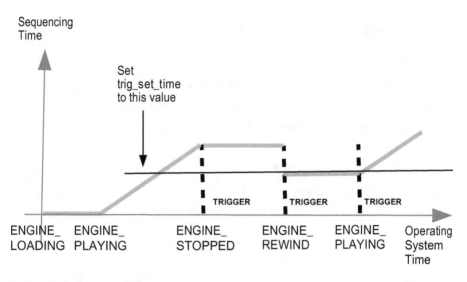

Figure 3-5. *Stop, then rewind*

Time Modules

The modules that handle all that timing are as follows:

- system → system_sequencer_control

- system → time

With the system_sequencer_control module, you can trigger start, stop, and rewind events and set trig_set_time. With the time module, you can access both timers if you need them, meaning the operating system timer and the sequencing timer. Note that a couple of modules use the sequencing time internally, namely, the oscillators and some others, so even if you don't use the time module, the sequencing timer might affect your rendering pipeline nevertheless.

Summary

In this chapter, you learned about the two timing concepts used in ThMAD, namely, the operating system time and the sequencing time. You investigated what they mean and how you can control them.

You also learned more about ThMAD meshes and ThMAD particle systems.

In the next chapter, you will learn about shaders, which allow for high-performance visualizations running almost completely on the graphics hardware.

CHAPTER 4

Shaders

This chapter covers shaders, which are a high-performance way of altering and generating 3D object data, including the position and coloring of vertices and surface fragments.

The shading constructs used in ThMAD 1.0 are based on an OpenGL version prior to 4.*x*. Quite a few of the functions and constructs used by ThMAD have now been marked as deprecated, which means they are subject to being removed in later versions. A future version of ThMAD will include an upgrade of both the OpenGL version and the way shading gets addressed. For now, you will have to go with this older version of shading.

Introduction to Shaders

Shaders were originally used to control various coloring aspects, which is where their name comes from. Later the concept was extended to control more objects inside the rendering pipeline. Today, there are a couple of different shader types that directly run on the graphics hardware to operate on different object types.

- Vertex shaders receive vertex data and allow for transformations such as from a 3D vertex to a 2D-object with depth.

```
Vertex-3D → Vertex -2D + Depth
```

© Peter Späth 2018
P. Späth, *Advanced Audio Visualization Using ThMAD*,
https://doi.org/10.1007/978-1-4842-3504-1_4

They describe the projection of the incoming 3D data onto a 2D plane for further processing. This not only allows for controlling positional vertex coordinates but also allows for other attributes pinned to vertices, namely, colors, normals, texture coordinates, and fog depth coordinates. In addition, you can add custom user-defined vertex attributes. Since nothing is hindering you from defining that 3D to 2D mapping in a non-Euclidian way, the number of interesting visualization effects is potentially endless.

- Geometry shaders are a relatively new type of shader that not all graphics cards and the more recent OpenGL versions can handle. They act on geometric objects built of several vertices at once and also allow for generating new graphics objects. Geometry shaders come right after the vertex shaders in the rendering pipeline.

- Tesselation shaders are also relatively new, and not all graphics cards and OpenGL versions can handle them. Tesselation allows you to fine-tune the process of subdividing surface elements into smaller pieces for improved rendering quality and improved performance.

- Fragment or pixel shaders allow you to directly control the outgoing pixel colors. They come in at a later stage of the rendering pipeline where the vertex information is no longer available. Since, however, a shader may access pixel colors at other coordinates, you can achieve various effects using this type of shader.

Shaders use their own C-like language, which must be compiled and then sent to the graphics card. Rendering engines in general and ThMAD specifically provide for shader program editing capabilities and do the rest for you.

ThMAD currently handles the vertex and fragment (pixel) shaders. Addressing other shaders may be included in future versions of ThMAD.

Vertex Shaders in Depth

Vertex shaders receive vertex positional coordinates and additional vertex attributes one at a time. That is, for each vertex received, the associated vertex shader program receives information only for a single vertex and does not see any of the other vertices.

The shader program, however, also receives so-called uniform variables, in short *uniforms*, which are common to all vertices and can be changed dynamically from outside the shader. See Figure 4-1.

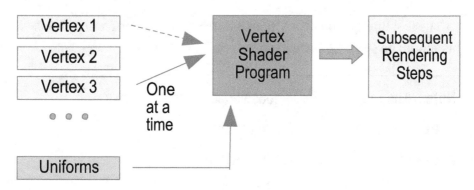

Figure 4-1. *Vertex shaders*

Simple Vertex Shaders

A simple vertex shader program looks like this:

```
#version 130
void main(void)
{
    vec4 pos = gl_Vertex;
    gl_Position =
        gl_ModelViewProjectionMatrix * pos;
}
```

Here is a shortcut for the program:

```
#version 130
void main(void)
{
    gl_Position = ftransform();
}
```

The first line specifies the OpenGL version in use. For ThMAD 1.0.0, this should be set to 130, which corresponds to OpenGL version 3.0.

Note   The mappings for various versions are as follows:

#version 110: OpenGL 2.0 from 2004

#version 120: OpenGL 2.1 from 2006

#version 130: OpenGL 3.0 from 2008

#version 140: OpenGL 3.1 from 2009

#version 150: OpenGL 3.2 from 2009

#version 330: OpenGL 3.3 from 2010

Which version is applicable depends on your graphics card.

The gl_Vertex variable is named by convention and contains the *homogeneous* positional model coordinates of the vertex in a four-dimensional vector. The first three coordinates stand for x, y, and z, and the fourth is usually set to 1.0.

The variable gl_ModelViewProjectionMatrix is another variable that gets filled in automatically for you. It contains the matrix product of the modelview matrix and the projection matrix and thus contains all the transformations you did on objects and the calculation according to the camera position and its type.

If you don't want to do anything fancy in the vertex shader or first need the transformed coordinates as an outcome from this matrix multiplication, using the function ftransform() does this but in a more concise manner.

The GLSL specification is an exhaustive reference of the GLSL language used for shaders. You can find it here:

```
https://www.khronos.org/registry/OpenGL/specs/gl/
GLSLangSpec.1.30.pdf
```

For this book and because of what is supported in ThMAD, you won't be using too many of the features available for shaders; you'll use just enough to allow for interesting visualizations.

Vertex Shader Variables

To get started with what you can actually do in ThMAD, in this section I will discuss the basic incoming and outgoing variables and describe control functions and other control constructs so that you can use the shaders for visualizations. Table 4-1 describes the most important variable types. You can look up all the other types in the specification.

Table 4-1. *Shader Variable Types*

Name	Description
bool	A Boolean describing a true or false condition. Example: bool isBigger = a > b;
int	A signed integer. Use uint instead if you need an unsigned integer. Example: int a = 7;
float	A floating-point number. Contrary to C, C++, and Java, there is no distinction between single- and double-precision floats. Example: float x1 = 7.3;
vecN	A floating-point vector with N components, where N can be one of 2, 3, or 4. To make a vector, write vecN name = vecN(coord1, coord2, ...); with as many arguments as necessary for the length specified. Example: vec4 myPos = vec4(0.0, 1.2, 0.0, 1.0);
matM	A floating-point matrix. M can be one of the following: <ul><li>2 or 2×2, which is a 2×2 matrix</li><li>3 or 3×3, which is a 3×3 matrix</li><li>4 or 4×4, which is a 4×4 matrix</li><li>2×3, 2×3, 2×4, 3×2,3×4, 4×2, or 4×3, which is an M1×M2 matrix</li></ul> For the initialization of matrices, you will use something like this: matM = matM(a11,...,a1M, a21, ..., a2M, ...) or this: matM = matM(vecM1, vecM2, ...) Example: mat2 m = mat2(0.0, 1.0, 0.1, 0.7);
sampler2D	A handle for 2D textures. You don't define textures in the shader code, but with their definition in the CPU program code (ThMAD), the textures as handled by the shaders will have this type.

For the input and output variables, see Table 4-2.

Table 4-2. *Vertex Shader In and Out Variables*

Class	Name	Description
In	gl_Vertex	The incoming vertex coordinates as a homogeneous coordinate set of type vec4.
	gl_Normal	The normal vector of the vertex; homogeneous coordinates of type vec4.
	gl_Color	The color assigned to the vertex; a vec4 typed variable with coordinates (RED, GREEN, BLUE, ALPHA).
	gl_MultiTexCoordN	Texture coordinates of each texture unit with N set to one of 0, 1, …, 7. ThMAD doesn't handle several texture units, so always use the first one, which is gl_MultiTexCoord0. The coordinates are of type vec4, with likely only the first two used.
	gl_FogCoord	A fog coordinate as a single float. ThMAD currently doesn't handle fog this way (you can misuse other variables for that).
	gl_ModelViewProjectionMatrix	A combination of three transformation matrices: the model matrix, the view matrix, and the projection matrix.
	[Attribute]	The specification allows for user-defined attributes added to each vertex. This is currently experimental in ThMAD, and you should not use it.

(continued)

51

Table 4-2. (*continued*)

Class	Name	Description
	[Uniform]	Uniforms are control-point variables common to all vertices and adjustable from outside the shaders. You must declare uniforms in the header of the program, as in uniform float myVariable;. ThMAD automatically provides an anchor for each uniform you specify. Uniforms other than those of type float are not supported in ThMAD.
Out	gl_Position	The transformed vertex coordinate of type vec4.
	gl_TexCoord[ ]	This is an array where texture coordinates get stored to pass them over to subsequent rendering stages; see the "Textures in Vertex Shaders" section.

Note   Many of these are marked as deprecated in the GLSL specification version 1.3. Future versions of ThMAD will use more modern constructs.

Operators and Functions

Table 4-3 describes which operators are available in shaders. Table 4-4 describes which functions are available in shaders.

Table 4-3. *Operators in Shaders*

Operator	Description
++ --	Increment and decrement. Can be placed before or after a variable. If placed before, it will be decremented or incremented before the variable gets used, otherwise afterward.
+ - ~ !	If placed before a variable, represents a positive or negative sign, a bitwise NOT, or a logical NOT.
* / %	Multiply, divide, remainder (modulus).
+ -	Plus, minus.
<< >>	Bit-wise shift.
< > <= >=	Relation; yields a Boolean value.
== !==	Equality, inequality; both yield a Boolean value.
& ^ |	Bit-wise AND, EXCLUSIVE OR, OR.
&& ^^ ||	Logical AND, EXCLUSIVE OR, OR.
cond. ? x : y	Selection.
+= -= /= *= %= <<= >>= &= ^= |=	In-place modifications.

Arithmetic Assignments

Using operators with different operand types quite often does what you'd expect. You can, for example, multiply a vector by a float, and what you will get is another vector with all its components multiplied by that number.

Table 4-4. *Functions in Shaders*

Function	Description
Angle and Trigonometry	
radians(x)	Converts the radian argument to degrees.
degrees(x)	Converts the degree argument to radians.
sin(a)	The sine function.
cos(a)	The cosine function.
tan(a)	The tangent function.
asin(a)	The (inverse) arcus sine.
acos(a)	The (inverse) arcus cosine.
atan(a)	The (inverse) arcus tangens.
atan(a,b)	The arcus tangent, given x and y coordinates. Will give the true angle, which the atan with just one argument can't. This is because x/-y is the same as -x/y.
sinh(x)	The hyperbolic sine.
cosh(x)	The hyperbolic cosine.
tanh(x)	The hyperbolic tangent.
asinh(x)	The inverse hyperbolic sine.
acosh(x)	The inverse hyperbolic cosine.
atanh(x)	The inverse hyperbolic tangent.
Exponential Functions	
pow(x,y)	The power function.
exp(x)	The exponential.
log(x)	The natural logarithm.

(continued)

Table 4-4. (*continued*)

Function	Description
`exp2(x)`	The 2^x function.
`log2(x)`	The logarithm to basis 2.
`sqrt(x)`	The square root.
`inversesqrt(x)`	Equals 1 / square root.
Common Functions	
`abs(x)`	Removes the sign.
`sign(x)`	Gives -1 for negative numbers, 0 for 0, otherwise +1.
`floor(x)`	Rounds down to the next integer.
`trunc(x)`	If x > 0, returns floor(x), otherwise ceil(x).
`round(x)`	Returns the integer nearest to x. If the fraction of x is 0.5, whether to round down or up is undefined.
`roundEven(x)`	Same as round; for a fraction of 0.5, the nearest even integer gets chosen.
`ceil(x)`	Rounds up to the next integer.
`fract(x)`	Gives the fraction; same as x – floor(x). Beware that fract(-3.2) thus yields 0.8.
`mod(x,y)`	Gives the modulus: x - y * floor(x/y).
`modf(x,y)`	Returns the fractional part of x and writes the integer part into y (as an out parameter). Both the result and y will have the same sign as x.
`min(x,y) max(x,y)`	The minimum or maximum.
`clamp(x,min,max)`	Clamps x if it gets smaller than min or greater than max.
`mix(x,y,a)`	Interpolates and yields x * (1-a) + y * a.

(*continued*)

Table 4-4. (*continued*)

Function	Description
step(x,e)	Gives 0.0 if x < e, else 1.0.
smoothstep(e0,e1,x)	Gives 0.0 if x < e0 or 1.0 if x > e1. Otherwise, the Hermite interpolation.
isnan(x) isinf(x)	Returns true if x is NaN (not a number) or infinite.
Geometric Functions	
length(v)	The Euclidian length of a vector.
distance(v1,v2)	The Euclidian distance between two points.
dot(v1,v2)	The dot product of two vectors.
cross(v1,v2)	The cross product of two vec3.
normalize(v)	Normalizes a vector, in other words, sets its length to 1.0.
ftransform()	Only for vertex shaders: the position of the vertex after multiplication with the model, view, and projection matrices.
faceforward(N,I, Nref)	With dot(Nref, I) < 0, returns N, otherwise -N. I is the incident vector, N the face normal.
reflect(I,N)	With I the incident vector and N the face normal, returns the reflection direction I – 2*dot(N,I) * N.
refract(I,N,e)	For the normalized incident vector I and the normalized surface normal N and the refraction ratio e, calculates the refraction vector.

(*continued*)

Table 4-4. (*continued*)

Function	Description
Matrix Functions	
`matrixCompMult(m1,m2)`	Multiplies matrix elements element-wise.
`outerProduct(v1,v2)`	The outer product of two vectors, res(i,j) = v1(i) * v2(j).
`transpose(m)`	Transposes a matrix.
Vectors	
`lessThan(v1,v2)` `lessThanEqual(v1,v2)` `greaterThan(v1,v2)` `greaterThanEqual(v1,v2)` `equal(v1,v2)` `notEqual(v1,v2)`	Compoent-wise comparison.
`any(bv) all(bv)` `not(bv)`	Boolean vector functions: "any" returns true if any vector component is true, "all" returns true if all vector components are true, "not" inverts true/false for all vector components.
Noise Functions	
`noise1(x)`	Generates a random float from [-1.0;1.0].
	The distribution is Gaussian. The output will be repeatable. That is, for a given fixed x, the output will be the same. You can pass in the operating system time as a *uniform* to feed x.
	WARNING: Not implemented in every graphics card; you should not use it. See the "Randomness in Shaders" section for how to include randomness.

(*continued*)

Table 4-4. (*continued*)

Function	Description
noise2(x)	Outputs a vec2 (members independent).
	WARNING: Not implemented in every graphics card; you should not use it. See the "Randomness in Shaders" section for how to include randomness.
noise3(x)	Outputs a vec3 (members independent).
	WARNING: Not implemented in every graphics card; you should not use it. See the "Randomness in Shaders" section for how to include randomness.
noise4(x)	Outputs a vec4 (members independent).
	WARNING: Not implemented in every graphics card; you should not use it. See the "Randomness in Shaders" section for how to include randomness.
Conversion	
int(bool)	Converts a bool to an int: false = 0, true = 1.
int(float)	Converts a float to an int.
float(bool)	Converts a bool to a float: false = 0.0, true = 1.0.
float(int)	Converts an int to a float.
bool(float)	Converts a float to a bool: 0.0 = false, else = true.
bool(int)	Converts an int to a bool: 0 = false, else = true.

Unless otherwise noted, if you apply functions primarily targeted at float values to vectors instead, the calculation will happen component-wise. For example, sqrt(x) applied to a vector will result in a vector with sqrt() applied to all the input vector's components. To check for each case in question, take a look at the specification.

Textures in Vertex Shaders

Textures (images that fill out the space between vertices) and bitmaps that are uploaded to the graphics hardware come into the vertex shaders as variables holding vec4 types (in other words, four-dimensional vectors).

- gl_MultiTexCoord0

- gl_MultiTexCoord1

- ...

- gl_MultiTexCoord7

For two-dimensional textures that are used by ThMAD, only the first two coordinates get used, and they specify the s and t coordinates of the texture (for textures, you write s and t instead of x and y), with the number range of [0;1].

For them to be transported to the next rendering step, you use the array glTexCoord[ ]. The elements of this array correspond to different texture units, but ThMAD can handle only the first and also only provides for gl_MultiTexCoord0. So to do the transport, write the following:

glTexCoord[0] = gl_MultiTexCoord0;

Since both are of type vec4, you can also introduce some coordinate manipulation here before the assignment to glTexCoord[0].

Fragment Shaders in Depth

Fragment shaders handle the coloring of pixels. Inside the graphics hardware's rendering pipeline, the translation from vertices to screen pixel coordinates has been performed in this stage. Not only do the corners or edges of figures arrive at the fragment shaders, but also all the pixels making up the surface parts do.

The vertex points themself as identifiable items are lost here, so you don't know to which vertices the pixels belong. However the graphics hardware calculated interpolation values for colors, normals, and matching texture coordinates to be applied at each pixel.

Simple Fragment Shaders

The simplest fragment shader is as follows:

```
#version 130
uniform sampler2D sampler;

void main() {
  vec4 tex =
    texture2D(sampler, gl_TexCoord[0].st);
  gl_FragColor =
    vec4(tex.r, tex.g, tex.b, tex.a);
}
```

The first line is the shading language version again, and the second line connects to a shader definition from inside the rendering engine of ThMAD. The name sampler is not given by convention; it is the name given by ThMAD to the texture object.

The texture2D() function contains all the magic that is done in the rendering pipeline up to that stage. As a first argument, it takes the texture object, and as a second argument, it takes the s / t (or x / y) coordinates from the texture. The gl_TexCoord[0] function that you have already seen in the vertex shaders will at this stage have a different content; all of its coordinates now have been transformed to pixel coordinates, so you don't have to deal with the vector transformations inside the fragment shader. Everything has been done for you. The gl_FragColor function is named by convention and signifies the color output of the fragment shader as

(RED, GREEN, BLUE, ALPHA). Here it just copies the output from the texture calculation to the output of the shader.

By the way, ".st" is a trick available to the shading language. The gl_TexCoord[0] function is a four-dimensional vector, but for the texture2D() function you need a two-dimensional vector. The ".st" trick just takes the first two elements and does the conversion you need here.

The second argument to texture2D() is also the place where you can do special calculations. By altering the coordinates here, you can do interesting things and also address pixels from the vicinity or from anywhere.

Fragment Shader Variables

Table 4-5 lists the most important variables for use inside fragment shaders. For others and for more details, please see the specification.

Table 4-5. *Fragment Shader In and Out Variables*

Class	Name	Description
In	gl_TexCoord[]	Interpolated texture coordinates as an array of vec4 vectors. If you have just one texture as in ThMAD, you use the first element of this array: gl_TexCoord[0].
	gl_FragCoord	The window-relative coordinates (x, y, z, 1/w) of the current fragment, a vec4 vector. The first two members, gl_FragCoord.x and gl_FragCoord.y, are pixel screen coordinates. The .z member is the fragment's depth mapped onto [0;1]. The fourth coordinate is the perspective division 1/w.

(continued)

Table 4-5. (*continued*)

Class	Name	Description
	[Uniform]	Uniforms are control-point variables common to all invocations of the vertex shader and adjustable from outside, in other words, from the ThMAD state. You must declare uniforms in the header of the program like `uniform float myVariable;`. ThMAD automatically provides an anchor for each uniform you specify. Uniforms other than those of type `float` are not supported in ThMAD.
Out	gl_ FragColor	The pixel's output color of type vec4: (RED, GREEN, BLUE, ALPHA), all in the range [0;1].

Note   Many of these are marked as deprecated in the GLSL specification version 1.3. Future versions of ThMAD will use more modern constructs.

Fragment Shader Operators and Functions

Unless otherwise noted, you can use the same operators and functions as for vertex shaders.

Varying Variables

Varying variables are special constructs that you define in the vertex shader but are then transformed and transported to the fragment shader. The *transformation* here means an interpolation; say, for example, you have triangle built of three vertices v1, v2, and v3, and the numbers v1 → 1.0, v2 → 3.0, v3 → 2.0 assigned to them inside the vertex shader.

```
varying float v;
// .. rest of the vertex shader, setting v
```

Now with the same declaration inside the fragment shader, the value of v will be interpolated.

```
varying float v;
// .. rest of the fragment shader
```

In other words, it will be 1.0 at vertex v1, 3.0 at v2, and 2.0 at v3. But in between, a smooth transition will be calculated, as shown in Figure 4-2 (black is for v = 0, white is for v = 3).

Figure 4-2. *Interpolation of varying variables*

Textures in Fragment Shaders

The most prominent texture-related function you will use is the texture2D( ) function (Table 4-6). For all other texture functions, please see the specification.

Table 4-6. *Texture-Related Functions in Fragment Shaders*

Function	Description
Textures	
`texture2D(sampler,v)`	Calculates the color value at a position v using the texture sampler. The texture must have been declared in the fragment shader code header as follows: `uniform sampler2D sampler;` The mapping of `sampler` to a texture will be done by ThMAD for you.

Advanced Lighting in Fragment Shaders

Both lights and material parameters are accessible from the shaders. The variable array `gl_LightSource[]` provides for all the lights registered. Each element is a struct, as shown here:

```
{
    vec4 ambient;             // Color, Aclarri
    vec4 diffuse;             // Color, Dcli
    vec4 specular;            // Color, Scli
    vec4 position;            // Vector, Ppli
    vec4 halfVector;
    vec3 spotDirection;       // Vector, Sdli
    float spotExponent;       // Srli
    float spotCutoff;         // Crli
                              // (range: [0.0,90.0], 180.0)
    float spotCosCutoff;      // Derived: cos(Crli)
                              // (range: [1.0,0.0],-1.0)
    float constantAttenuation;    // K0
    float linearAttenuation;      // K1
    float quadraticAttenuation;   // K2
}
```

To get, for example, the diffuse color of the first light, you'd write the following:

```
gl_LightSource[0].diffuse
```

Lighting is explained in detail in the OpenGL specification, available here:

```
https://www.khronos.org/registry/OpenGL/specs/gl/glspec30.pdf
```

Most important are, of course, the position and the color values for ambient, diffuse, and specular lights. For specular lights, you are also interested in the exponent and in the cutoff angle.

For the material, you provide `gl_FrontMaterial` and `gl_BackMaterial`, and again you have structs with the following contents:

```
{
    vec4 emission;      // Color, Ecm
    vec4 ambient;       // Color, Acm
    vec4 diffuse;       // Color, Dcm
    vec4 specular;      // Color, Scm
    float shininess;    // Float, Srm
}
```

To access, for example, the diffuse color of the material's front side, you'd write the following inside the shader:

```
gl_FrontMaterial.diffuse
```

How you use all these variables depends on what you want to achieve. Of course, you can build up lots of unrealistic lighting scenes from them, but there are a couple of realistic color models as well. One of the best known realistic models is the Phong illumination model, and to implement it, you'd use the following for the vertex shader:

```
#version 130
varying vec3 N;
varying vec3 v;

void main(void) {
    v = vec3(gl_ModelViewMatrix * gl_Vertex);
    N = normalize(gl_NormalMatrix * gl_Normal);

    gl_Position = gl_ModelViewProjectionMatrix *
          gl_Vertex;

}
```

You'd use the following for the fragment shader:

```
#version 130
varying vec3 N;
varying vec3 v;
void main (void) {
    vec3 L = normalize(gl_LightSource[0].position.xyz
        - v);
    vec3 E = normalize(-v); // we are in Eye
                            // Coordinates, so EyePos
                            // is (0,0,0)
    vec3 R = normalize(-reflect(L,N));

    //calculate Ambient Term:
    vec4 Iamb = gl_FrontLightProduct[0].ambient;

    //calculate Diffuse Term:
    vec4 Idiff = gl_FrontLightProduct[0].diffuse *
          max(dot(N,L), 0.0);
    Idiff = clamp(Idiff, 0.0, 1.0);
```

```
// calculate Specular Term:
vec4 Ispec = gl_FrontLightProduct[0].specular *
     pow(max(dot(R,E),0.0),
          0.3*gl_FrontMaterial.shininess);
Ispec = clamp(Ispec, 0.0, 1.0);

// write Total Color:
gl_FragColor =
     gl_FrontLightModelProduct.sceneColor
     + Iamb + Idiff + Ispec;
}
```

This is without textures.

Note   These shaders are available under B-3.6_Shader_
Lighting in the TheArtOfAudiovisualization folder.

These are the steps to follow for this example:

1. Take a simple vertex shader, but add as varying
    variables the normal vector N and the position v in
    the modelview space, that is, without the camera's
    (or eye's) position taken into account.

2. Now inside the fragment shader, calculate L as
    the normalized direction vector from the point in
    question to the light source. Because you're using
    v, this happens in the modelview space, that is,
    without the camera's (or eye's) position taken into
    account.

3. Calculate E as the normalized direction to the
    camera's position, assumed to carry (0,0,0)
    coordinates.

4. Calculate R as the normalized reflection direction for a perfect reflection of the light beam (which uses the reflect() function described earlier).

5. The rest is the actual Phong formula, described by the comments in the code. The gl_FrontLightProduct variable is a convenience variable that is calculated as the product of the front material colors and the corresponding lighting colors.

Figure 4-3 shows the calculated vectors (vectors not normalized there).

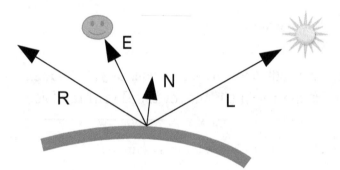

Figure 4-3. *Vectors inside the Phong illumination model*

Using Shaders from Inside ThMAD

A basic state for utilizing shaders in ThMAD consists of four modules.

- The output module: outputs → screen0

- The shader render module: renderers → shaders → glsl_loader

- A texture renderer, for example: renderers → basic → textured_rectangle

- A texture generator (or loader), for example: texture → particles → blob

Figure 4-4 shows the state.

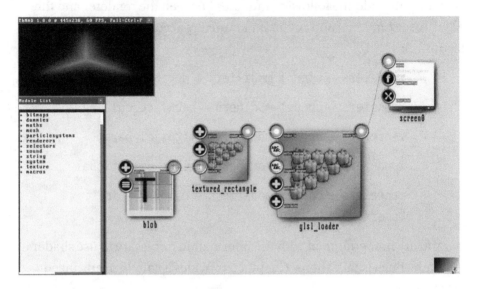

Figure 4-4. *Basic state for shaders*

If you want to avoid the texture being repeated, just put the texture →
modifiers → tex_parameters module between the texture generator and
the texture renderer and set its parameter accordingly.

To write the shader's code, open the shader editors of the glsl_loader
module. The editors have only limited capabilities, but you can use any
editor of your choice and transport the contents back and forth over the
system's clipboard.

As soon as you add any uniforms to your shaders, ThMAD will
automatically generate the corresponding subanchors inside the
"uniforms" anchor of the glsl_loader module, and you can connect those
subanchors to any control pipeline like you would for any other anchor.
Remember, those uniforms run directly on the graphics hardware, so you
can make some incredibly fast things here!

For the simplest shader state using material properties, lights, and a camera, you'd add the following modules between the renderer and the `glsl_loader` module and set their parameters accordingly, as shown in Figure 4-5:

- renderers → opengl_modifiers → depth_buffer

- renderers → opengl_modifiers → backface_culling

- renderers → opengl_modifiers → material_param

- renderers → opengl_modifiers → light_directional

- renderers → opengl_modifiers → cameras → orbit_ camera

You are now equipped with the information necessary to use shaders for interesting visualizations. Chapter 5 provides a couple of advanced examples.

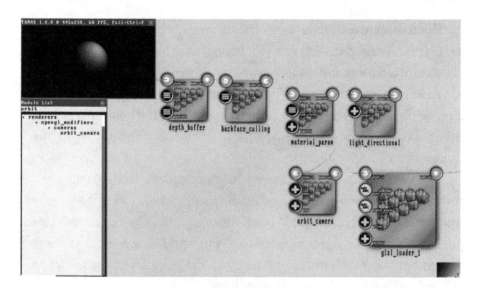

Figure 4-5. *State for shaders with lighting*

Randomness in Shaders

To get randomness in shaders, frequently using a noise texture is suggested. Unfortunately, this is not an easy approach for ThMAD since if you also need image textures, you have a problem: ThMAD allows for only one texture in a shader. There are several approaches, though, for algorithmic randomness. Note that in reality you have only pseudorandomness. The generated numbers look random and in most cases may behave like they are purely random, but in fact they cover some hidden nonrandom patterns.

A lot of ideas are being realized when it comes to randomness in shaders. However, for these purposes, I present only the basic functions here for creating a float random number given a float, a vec2, or a vec3 type input:

```
float rand(float n){
    return fract(sin(n) * 43758.5453123);
}

float rand(vec2 n) {
     return fract(sin(dot(n, vec2(12.9898, 4.1414)))
                  * 43758.5453);
}

float noise(float p){
    float fl = floor(p);
    float fc = fract(p);
    return mix(rand(fl), rand(fl + 1.0), fc);
}

float noise(vec2 n) {
  const vec2 d = vec2(0.0, 1.0);
  vec2 b = floor(n),
       f = smoothstep(vec2(0.0),vec2(1.0), fract(n));
```

```
  return mix(
      mix(rand(b), rand(b + d.yx), f.x),
      mix(rand(b + d.xy), rand(b + d.yy), f.x), f.y);
}

float mod289(float x){return x -
    floor(x * (1.0 / 289.0)) * 289.0;}
vec4 mod289(vec4 x){return x -
    floor(x * (1.0 / 289.0)) * 289.0;}
vec4 perm(vec4 x){return
    mod289(((x * 34.0) + 1.0) * x);}

float noise(vec3 p){
    vec3 a = floor(p);
    vec3 d = p - a;
    d = d * d * (3.0 - 2.0 * d);

    vec4 b = a.xxyy + vec4(0.0, 1.0, 0.0, 1.0);
    vec4 k1 = perm(b.xyxy);
    vec4 k2 = perm(k1.xyxy + b.zzww);

    vec4 c = k2 + a.zzzz;
    vec4 k3 = perm(c);
    vec4 k4 = perm(c + 1.0);

    vec4 o1 = fract(k3 * (1.0 / 41.0));
    vec4 o2 = fract(k4 * (1.0 / 41.0));

    vec4 o3 = o2 * d.z + o1 * (1.0 - d.z);
    vec2 o4 = o3.yw * d.x + o3.xz * (1.0 - d.x);

    return o4.y * d.y + o4.x * (1.0 - d.y);
}
```

The noise() function will then produce the random number. As input, you can, for example, use the coordinates of vertices or the fragment position vector available to the vertex and fragment shaders, respectively. You can find more about random numbers and noise in shaders if you search the Internet for *shader random* or *shader noise* using your favorite search engine.

Summary

After learning about vertex and fragment shaders and their functioning inside the OpenGL graphics pipeline, you learned how to use shaders from within ThMAD.

In the next chapters, you will be looking at advanced stories and seeing how to unleash the full power of ThMAD for your own ideas.

CHAPTER 5

Stories I

This chapter presents advanced example states and is an extension to the stories from the book *Audio Visualization Using ThMAD*, so once in a while it will refer to the stories presented there. The stories in this chapter are somewhat self-contained, though, so you do not need to have that other book on hand, although it is recommended as a reference.

Textures Revisited

In *Audio Visualization Using ThMAD*, you saw how to deal with textures, that is, images or in general bitmap data stored on the graphics hardware. This chapter will extend the ideas presented there and show some more techniques for how to use textures.

Texture Distortion via Bitmaps

Note This sample is available under `B-5.1_Texture_bitmap_distortion` in the `TheArtOfAudioVisualization` folder.

Moving on from what you already learned about texture coordinate distortion in *Audio Visualization Using ThMAD*, there is another module for texture distortion using a bitmap.

- mesh → texture → mesh_tex_bitmap_distort

© Peter Späth 2018
P. Späth, *Advanced Audio Visualization Using ThMAD*,
https://doi.org/10.1007/978-1-4842-3504-1_5

It works as follows: the usual texture coordinate space in the u-v plane $(0,0) \rightarrow (1,1)$ first gets mapped to a scaled and translated version, as shown here:

```
uᵢ → uᵢ' = u_scale · uᵢ + u_translate
vᵢ → vᵢ' = v_scale · vᵢ + v_translate
```

Here, i iterates over all texture coordinates. With these coordinates, you apply a bitmap function, as shown here:

```
( ui', vi') → ( ui'', vi'') = ( ui', vi') + -1 + 2 · bitmap(ui', vi')
            for both ui' and vi' inside [0;1] or else
            0
```

Here, `bitmap(q, r)` is the (RED, GREEN) tuple at position (q · WIDTH, r · HEIGHT) of the bitmap. For the purposes of this chapter, both RED and GREEN have values from [0;1]. So, the RED and GREEN pixel values of the bitmap describe the extra distortion, and you can use a drawing program to actually *paint* the desired distortion. Because of the `-1 + 2 · bitmap(uᵢ', vᵢ')` in the previous formula, you have *no* distortion for the RED and GREEN values, which are both 0.5. So, you can reduce the texture coordinate for bitmap(...) to less than 0.5 and increase it for bitmap(...) to greater than 0.5.

Next, you roll back the first transformation, as shown here:

```
uᵢ'' → uᵢ''' = [ uᵢ'' - u_translate ] / u_scale
vᵢ'' → vᵢ''' = [ vᵢ'' - v_translate ] / v_scale
```

Then you replace the original texture coordinates with those you received last.

The module mesh_tex_bitmap_distort therefore has anchors for the mesh for u_scale, v_scale, u_translate, v_translate, and the input bitmap. In addition, you can use the intensity anchor to control the intensity effect of the bitmap; 0 means none at all, and 1 means 100 percent intensity.

As an example, see Figure 5-1. The modules not shown in the figure include the usual 3D pipeline (light, material, camera and so on).

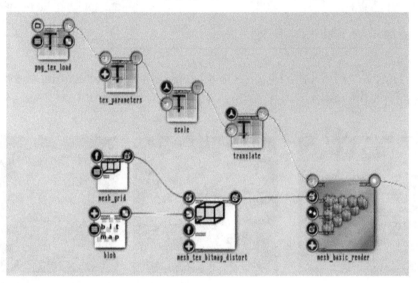

Figure 5-1. *Bitmap texture distortion*

Tables 5-1 and 5-2 list the anchor values.

Table 5-1. *Blob Parameters*

bitmaps → generators → blob	
settings/arms	5.0
settings/attenuation	0.8

Table 5-2. *Mesh Tex Bitmap Distort Parameters*

mesh → texture → mesh_tex_bitmap_distort		
intensity	0.1	Rate of distortion; here 10.0 percent
spatial/u_off	0	Offset before applying the bitmap
spatial/v_off	0	
spatial/u_scale	0.64	Scale before applying the bitmap
spatial/v_scale	0.88	

Figure 5-2 shows the results. On the left is the original texture, and on the right is the bitmap used for distortion.

Figure 5-2. *Bitmap texture distortion output*

Using Shaders

Shaders are programs that run directly on the graphics hardware. They are something like the holy grail of computer game development, but they also have gained some considerable attention for high-performance computing since shader programs can be executed highly parallelized.

Shaders can access vertex data, pixel data, and texture data and thus allow for elaborate graphics operations. For now you will concentrate on using shaders for texture mapping.

You will focus on two shader types: the vertex shader and the fragment shader. There are more, but the others are special extensions for certain types of graphics cards or cover corner cases you do not need right now. See Chapter 4 for an introduction on shaders.

The vertex shader gets vertex coordinates, one vertex at a time. Its sole purpose is to give back a calculated coordinate for the vertex it has as input. Geometrical transformations such as rotation, scaling, skewing, translation, or distortion of one or another kind can be done using vertex shaders. A vertex shader can add additional information as well; quite often a vertex shader is used to prepare for using a texture for subsequent rendering steps. It can do this because it has access to textures that are registered in the rendering process.

A basic vertex shader looks like this:

```
void main(void)
{
    gl_TexCoord[0] = gl_MultiTexCoord0;
    gl_Position = gl_ModelViewProjectionMatrix
        * gl_Vertex;
}
```

It first assigns the multitexture number zero to an internal array gl_TexCoord[0]. Multitextures are beyond the scope of the current purpose, but the zero accesses the texture you will be assigning to the rendering process using ThMAD. Subsequent rendering subprocesses can then refer to the texture by accessing gl_TexCoord[0]. The other line, gl_Position = ..., calculates the output of the shader; the name gl_Position is predefined and cannot be changed. The formula behind the equal sign uses the standard way to project the input gl_Vertex, which holds the incoming vertex coordinates. This vertex shader program does nothing that the renderer would do if no shader program was provided. It, however, gives a starting point because various other calculations can be introduced before presenting the final value to gl_Position.

The fragment shader gets invoked on a per-pixel basis. It can be used to receive and transform color values, and it can be used to add texture data to the output in quite a free manner. It can even misuse incoming color or texture data to do something completely different, which can lead to some highly interesting and extremely high-performing effects. The standard for texture mapping is, however, expressed by the following basic fragment shader code:

```
uniform sampler2D
sampler; void main() {
  vec4 tex = texture2D(sampler, gl_TexCoord[0].st);
  gl_FragColor = vec4(tex.r, tex.g, tex.b, tex.a);
}
```

The `uniform sampler2D sampler;` code hides the internal prepreparation for the texture mapping. Consider it like an introductory step for interfering with the mapping. The `vec4 tex = ...` line actually determines the texture color after some linear default mapping. Since the second argument, `gl_TexCoord[0].st`, is a point in the texture space (0,0) → (1,1), you could prior to providing it to the `texture2D()` function save it in a variable and transform it in any way you want. You will see later in the chapter how to do that. The last line, `gl_FragColor = ...`, provides the pixel color as an output. This is actually the main responsibility of the fragment shader; you can actually do whatever you like, but there must be a line with `gl_FragColor = ...` at the end. The fragment shader code just presented does the same thing as if it didn't exist. Just as for the vertex shader, it is the starting point for your own attempts to interfere with the rendering process.

Let's discuss how you can use shaders in ThMAD. You start with a basic state just using the default shaders and then gradually introduce new features.

Note   The following samples are under B-5.1_Shaders_for_
textures* in the TheArtOfAudioVisualization folder.
The shader code is provided in the following folder: [ThMAD_
INST]/share/thmad/TheArtOfAudioVisualization-
snippets/B-5.1_Using_Shaders

First add the following modules to your canvas:

- texture → loaders → png_tex_load

- texture → modifiers → tex_parameters

- renderers → basic → textured_rectangle (twice)

- renderers → shaders → glsl_loader

Connect them as shown in Figure 5-3.

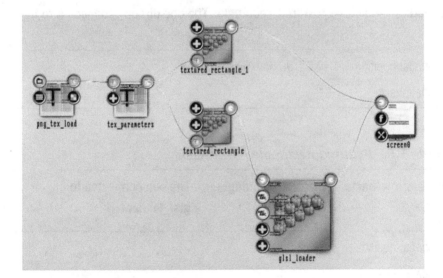

Figure 5-3.   *Using shaders for texture mapping*

For the parameters, see Tables 5-3 to 5-7.

Table 5-3. *Png Tex Load Parameters*

texture → loaders → png_tex_load		
file name	Any	Choose one from the resources folder.

Table 5-4. *Tex Parameters*

texture → modifiers → tex_parameters	
parameters/wrap_s	clamp
parameters/wrap_t	clamp

Table 5-5. *Textured Rectangle Parameters*

renderers → basic → textured_rectangle	**The one directly connected to the screen**
spatial/position	−0.75, 0.75, 0
spatial/size	0.2

Table 5-6. *Textured Rectangle Parameters*

renderers → basic → textured_rectangle	**The one connected to glsl_loader**
spatial/position	0, 0, 0
spatial/size	1.0

Table 5-7. *Glsl Loader Parameters*

renderers → shaders → glsl_loader	
vertex_program	Explained in the following text
fragment_program	Explained in the following text

You can put a texture as a PNG file in /home/ [USER]/thmad/*/data/ resources first. Set it to size 128×128 or 256×256 or 512×512. You might have to restart ThMAD to see any new file.

The module gets created with the default code shown previously. In the fragment shader code, replace the following:

```
gl_FragColor = vec4(tex.r, tex.g, tex.b, tex.a);
```

with this to see the shaders at work:

```
gl_FragColor = vec4(tex.r, 0.0, tex.b, tex.a);
```

To do that change, double-click the fragment_program anchor of glsl_loader, and inside the editor that then appears change the text accordingly. Clicking the Save button will upload the changes to the graphics hardware. The output should look like Figure 5-4 (on the top left, the original gets shown).

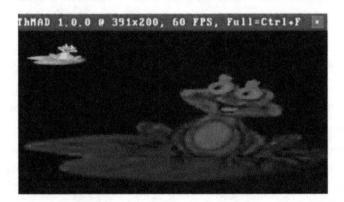

Figure 5-4. *Using shaders for colorizing texture*

83

This draws the texture with the green color channel muted. You can also swap color values or apply other mathematical operations to them. As another example, use the following in the fragment shader, which changes the color curves, effectively brightening the colors:

```
gl_FragColor = vec4(pow(tex.r,0.5),
    pow(tex.g,0.5),
    pow(tex.b,0.5), tex.a);
```

The output will look like Figure 5-5.

Note that the shader code editor has only limited editing capabilities, but you can use any text editor you like and then use the clipboard to transport code back and forth. That is what the To Clipboard and From Clipboard buttons are for.

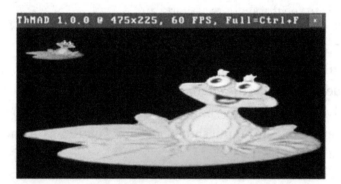

***Figure 5-5.** Using shaders for brightening textures*

To see a nontrivial shader coordinate mapping at work, you will try to construct a sample where one point "sucks up" the surrounding pixels, like in a black hole. You again use the fragment shader for that since the vertex shader knows only the four corners of the textured_rectangle module and you need much more fine-grained control over the coordinates. The new code is as follows (the line numbers are for display purposed only in the book):

```
(1)    uniform sampler2D sampler;
(2)    void main() {
(3)      vec2 suck_point = vec2(0.6, 0.3);
(4)      vec2 tc =  gl_TexCoord[0].st;
(5)      vec2 v =  tc - suck_point;
(6)      float d = clamp( length(v), 0.001, 1000.0);
(7)      vec2 distort = v * (0.2 / pow(d,1.0));
(8)      float ang = 0.1 / pow(d,1.0);
(9)      mat2 rot = mat2( cos(ang), -
             sin(ang), sin(ang), cos(ang)
             );
(10)     distort = rot * distort;
(11)     vec4 tex = texture2D ( sampler,
             tc + distort );
(12)     gl_FragColor = vec4(tex.r, tex.g,
             tex.b, tex.a);
(13)   }
```

Line 3 defines suck_point, line 4 gives you the calculated linear texture coordinates, and line 5 gives you the distance vector. At line 6 you calculate the numerical distance, but make sure it does not drop below 0.001. At line 7 you calculate a distance-related distortion, which decreases with the distance to suck_point. At line 8 you calculate a whirling angle that increases the nearer you get to suck_point, and line 9 deduces a rotation matrix from that. The same gets applied at line 10. You know the rest: you fetch the texture color value at the distorted point and assign it to the fragment shader output. The result will look like Figure 5-6.

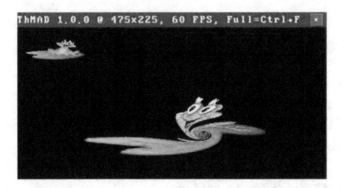

Figure 5-6. *Using shaders for texture coordinate distortion*

Wouldn't it be nice if you could control some parameters from the outside? Possible candidates are the distortion rates and the position of the suck_point value. It *is* possible, and this chapter will show you how. The procedure is as follows: at the top of the fragment shader code you can see the definition of a *uniform* variable. In this case, it is used for an internal field, but you can use uniform variables also to dynamically access the shader from the outside. You introduce uniforms for the suck_point position at line 3 and also for the distortion rates: 0.2 at line 7 for the distance distortion and 0.1 at line 8 for the swirl distortion. You thus get the following (line numbers are for display purposes only):

```
(1)    uniform sampler2D sampler;
(2)    uniform vec3 suck_point;
(3)    uniform float dist_distortion;
(4)    uniform float swirl_distortion;
(5)    void main() {
(6)      vec2 tc = gl_TexCoord[0].st;
(7)      vec2 v = tc -
                vec2(suck_point.x,
                suck_point.y);
(8)      float d = clamp( length(v), 0.001, 1000.0);
```

```
(9)      vec2 distort = v * (dist_distortion
             / pow(d,1.0));
(10)     float ang = swirl_distortion/pow(d,1.0);
(11)     mat2 rot = mat2( cos(ang), -
             sin(ang), sin(ang), cos(ang) );
(12)     distort = rot * distort;
(13)     vec4 tex = texture2D ( sampler, tc
             + distort );
(14)     gl_FragColor = vec4(tex.r, tex.g,
             tex.b, tex.a);
(15)   }
```

With the code changed accordingly and saved, ThMAD automatically adds new anchors to the `glsl_loader` module, as shown in Figure 5-7. Note that you have to make the `suck_point` value a three-dimensional vector because ThMAD doesn't know two-dimensional vectors. The third coordinate gets ignored. Go ahead and play with these values, or connect the anchors to sound input or oscillators. You can even achieve some rather surrealistic effects; see, for example, Figure 5-8 (`swirl_distortion` = 0.17, `dist_distortion` = -0.2, `suck_point` = (0.58; 0.50; 0.0)).

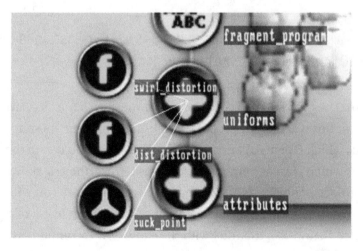

Figure 5-7. *New anchors due to custom shader parameters*

Figure 5-8. *Surrealistic shader mapping*

Explosions

Here you will take a look at how explosions can be achieved. We need an object with many parts which can be blown apart, and a sphere is a very good candidate here.

Exploding Star

Note   This sample is available under B-5-2_Explosion in the TheArtOfAudioVisualization folder.

ThMAD has a mesh_explode module for meshes. Explosions in real-world scenarios will blow apart the surface elements *and* the solid interiors of objects, which OpenGL cannot possibly do unless you create the interior of objects, which is an extremely intricate task. Nevertheless, blowing apart just the surface of objects is an interesting effect, and here you will see how to do that. More precisely, you will simulate the explosion of a planet in a science-fiction-like manner, with the explosion accompanied by a concentric ray burst.

For this explosion module, you first need a mesh with many parts. A good candidate is any sphere since spheres need many constituents to create a surface that seems to be round. For this planet explosion, you will use two interwoven spheres with different levels of detail to have exploded parts of two different sizes. To get started, right-click, select New, and then select Empty Project to create an empty canvas. Add the following modules:

- renderers → opengl_modifiers → cameras → orbit_camera
- renderers → opengl_modifiers → light_directional
- renderers → opengl_modifiers → auto_normalize
- renderers → opengl_modifiers → material_param
- renderers → opengl_modifiers → depth_buffer
- renderers → opengl_modifiers → gl_translate
- renderers → opengl_modifiers → gl_scale

Connect them with each other and to the screen0 module, as shown in Figure 5-9.

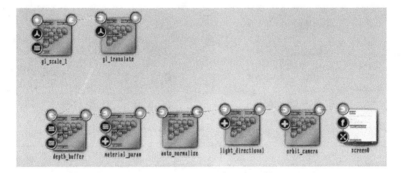

Figure 5-9. *Explosion, basic 3D setup*

The parameters are listed in Tables 5-8 to 5-14.

Table 5-8. *Screen0 Parameters*

screen0	
clear_color	0.02; 0.08; 0.15; 1.0

Table 5-9. *Orbit Camera Parameters*

renderers → opengl_modifiers → cameras → orbit_camera	
rotation	1.58; −0.70; 0.46
distance	15.1
upvector	0.75; −0.34; 0.56
fov	90.0
perspective_correct	yes

Table 5-10. *Light Directional Parameters*

renderers → opengl_modifiers → light_directional	
enabled	YES
position	0.29; 0.29; 0.12
ambient_color	0; 0; 0; 1
diffuse_color	0.58; 0.51; 0.89; 1.0
specular_color	0.97; 0.87; 0.87; 1.0

Table 5-11. *Material Param Parameters*

renderers → opengl_modifiers → material_param

ambient_reflectance	0.2; 0.2; 0.2; 1.0
diffuse_reflectance	0.32; 0.69; 0.86; 0.97
specular_reflectance	0.92; 0.92; 0.64; 0.94
emission_intensity	0.07; 0.16; 0.40; 1.0
specular_exponent	5.0

Table 5-12. *Depth Buffer Parameters*

renderers → opengl_modifiers → depth_buffer

depth_test	ENABLED
depth_mask	ENABLED

Table 5-13. *Gl Translate Parameters*

renderers → opengl_modifiers → gl_translate

translation	0; 0; 0

Table 5-14. *Gl Scale Parameters*

renderers → opengl_modifiers → gl_scale

scale	4.0; 4.0; 4.0	You can achieve surrealistic effects if you change this.

Next you will add the shapes and the explosion module.

- renderers → mesh → mesh_basic_render (twice)

- mesh → modifiers → deformers → mesh_explode (twice)

- mesh → solid → mesh_sphere_icosahedron (twice)

- maths → oscillators → oscillator

Connect them as shown in Figure 5-10.

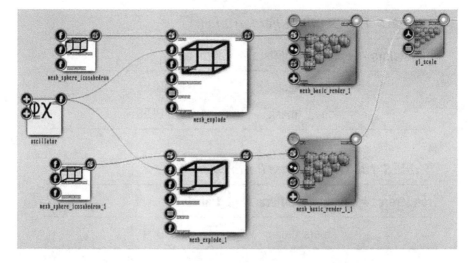

Figure 5-10. *Explosion, exploding shapes, and explosion modules*

Then enter parameters for them as shown in Tables 5-15 to 5-19.

Table 5-15. *Mesh Basic Render Parameters*

renderers → mesh → mesh_basic_render	Both
Leave all values at their defaults.	

Table 5-16. *Mesh Explode Parameters*

mesh → modifiers → deformers → mesh_explode		Both
start		Connected to the oscillator.
explosion_factor	1.8	You can change this to alter the explosion intensity. You can even choose different values for both modules.
velocity_deceleration	0.01	Particles slow down after the explosion. Change this value to alter the slowdown behavior.

Table 5-17. *Mesh Sphere Icosahedron Parameters*

mesh → solid → mesh_sphere_icosahedron		**First planet**
subdivision_level	6.0	
max_normalization_level	28.0	

Table 5-18. *Mesh Sphere Icosahedron Parameters*

mesh → solid → mesh_sphere_icosahedron		Second planet
subdivision_level	4.0	
max_normalization_level	10.0	

Table 5-19. *Oscillator Parameters*

Maths → oscillators → oscillator		Connected to the "start" trigger of both explosion modules
osc		
osc_type	square	
freq	0.14	
amp	3.0	The value does not matter, but amp + ofs must be > 0.
ofs	−0.1	
phase	0.0367	
options		
drive_type	time_internal_absolute	You need a global synchronization with other oscillators that you add later.

This is actually enough to see the explosion, as shown in Figure 5-11.

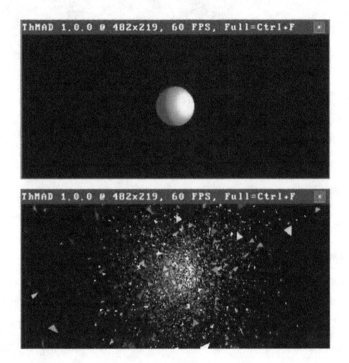

Figure 5-11. *Explosion, basic explosion output*

As an extra goody, you will add concentric burst beams that synchronize with the planet explosion. To do so, add these modules:

- renderers → opengl_modifiers → material_param
- renderers → opengl_modifiers → gl_scale_one
- renderers → opengl_modifiers → gl_rotate
- renderers → basic → textured_rectangle
- texture → loaders → bitmap2texture
- bitmap → generators → concentric_circles

Connect them as shown in Figure 5-12.

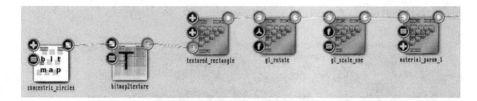

Figure 5-12. *Explosion, concentric beam base*

Set those modules' parameters to the settings listed in Tables 5-20 to 5-24.

Table 5-20. *Material Param Parameters*

renderers → opengl_modifiers → material_param		
ambient_reflectance	0.2; 0.2; 0.2; 1.0	
diffuse_reflectance	1; 1; 1; 1	You will later connect this to a visibility subpipeline to synchronize the circle's visibility with the explosion.
specular_reflectance	0; 0; 0; 1	
emission_intensity	1; 1; 1; 1	The beams are white.
specular_exponent	10.0	

Table 5-21. *Gl Scale One Parameters*

renderers → opengl_modifiers → gl_scale_one		
scale	7.0	You will later connect this to an oscillator.

Table 5-22. *Gl Rotate Parameters*

renderers → opengl_modifiers → gl_rotate	
axis	−0.058; 0.125; 0.935
angle	−0.23

Table 5-23. *Textured Rectangle Parameters*

renderers → basic → textured_rectangle	
spatial	
position	0.50; 0.52; 0.045
size	3.2

Table 5-24. *Concentric Circles Parameters*

bitmap → generators → concentric_circles		
frequency	1.0	Repetition frequency of the concentric circles.
attenuation	2.0	Circle or ring sharpness.
color	1; 1; 1; 1	White.
alpha	yes	Let the planets and planet parts shine through.
size	256x256	

You now need two additions. First, you want the concentric circles to inflate. Second, you want the concentric circles to disappear at a later stage of the explosion. Inflating will start from zero, so if you start the inflation at the correct instance of time, you will not see them until the explosion starts. But you must let them disappear later because there is no way to let them inflate "away." If you do not let them disappear later, you would see them hang around, which doesn't seem realistic.

Let's start with the first of the remaining tasks. Place the following on the canvas and connect them as shown in Figure 5-13:

- maths → arithmetics → ternary → float → mult_add
- maths → oscillators → float_sequencer
- maths → oscillators → oscillator
- maths → dummies → float_dummy
- maths → arithmetics → binary → float → div

The float dummy is connected to the freq anchor of the oscillator, and the oscillator is connected to the options / trigger anchor of the float_sequencer. Mult_add is connected to the scale anchor of gl_scale_one (not shown in Figure 5-13).

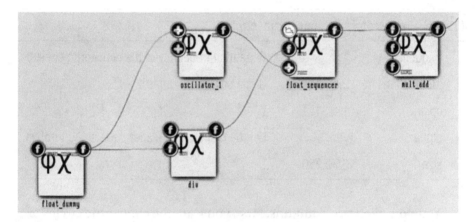

Figure 5-13. *Explosion, concentric rings control*

Set the parameters as shown in Tables 5-25 to 5-29.

Table 5-25. *Mult Add Parameters*

maths → arithmetics → ternary → float → mult_add	
Controls the inflation extent of the rings	
first_mult	15.0
then_add	0.0

Table 5-26. *Float Sequencer Parameters*

maths → oscillators → float_sequencer		
float_sequence	See the following text	
length	7.24	Later connected to the oscillator
options		
behavior	trigger	
time_source	operating_system	
trigger		Connected to the oscillator output
drive_type	time_internal_relative	

To define the float sequence for float_sequencer, open the control by double-clicking the anchor. The sequence input will then look like Figure 5-14.

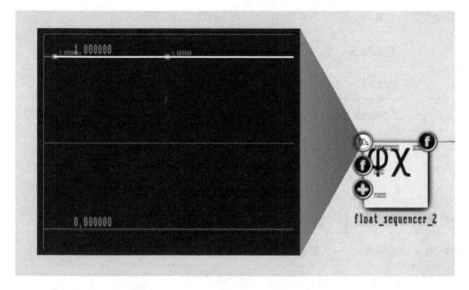

Figure 5-14. *Explosion, concentric rings control, float sequencer I default*

Make it look like Figure 5-15.

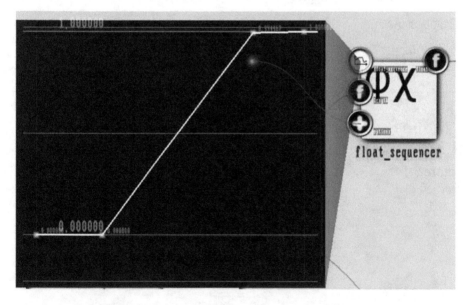

Figure 5-15. *Explosion, concentric rings control, float sequencer I*

Add anchors by holding Shift and clicking the line; then move them by dragging. This sequence makes sure the ring inflation will start at the correct time.

Table 5-27. *Oscillator Parameters*

maths → oscillators → oscillator		Controls the triggering of the sequencer module
osc		
osc_type	square	Square is the designated type for triggers.
freq		Controlled by the dummy below.
amp	0.5	Because of these two values, the
ofs	0.5	oscillator goes from 0.0 to 1.0 for triggering/resetting the trigger in the connected float sequencer.
phase	-0.543	Let the trigger start at the correct point in time.
options		
time_source	operating_system	
drive_type	time_internal_absolute	You need a global synchronization with other oscillators.

Table 5-28. *Float Dummy Parameters*

maths → dummies → float_dummy		Controls all oscillators
float_in	0.138	

Also connect the output from the float_dummy module to the freq anchor from the explosion subpipeline oscillator.

Table 5-29. *Div Parameters*

maths → arithmetics → binary → float → div	Controls all oscillators
param1 1.0	
param2	Connected to the float_dummy module; output will then be the period of an oscillation cycle.

This finishes the ring's inflation control. For the remaining rings' visibility control, place these modules on the canvas and connect them as shown in Figure 5-16:

- maths → converters → 4float_to_float4

- maths → oscillators → float_sequencer (another one)

- maths → oscillators → pulse_oscillator

The 4float_to_float4 module is connected to the anchor diffuse_reflectance of the material_param module from the concentric circle subpipeline. The pulse_oscillator module connects to the trigger anchor of the float sequencer. The modules float_dummy and div are from the subpipeline you already constructed.

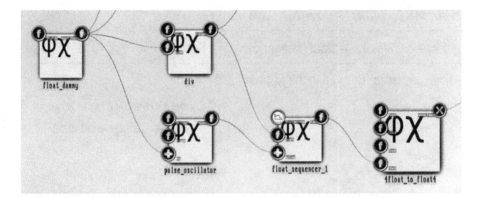

Figure 5-16. *Explosion, concentric rings visibility control*

Set the parameters as shown in Tables 5-30 to 5-31.

Table 5-30. *4Float To Float4 Parameters*

maths → converters → 4float_to_float4		Demultiplexes the color input
floata	1.0	You control only ALPHA; the rest is white.
floatb	1.0	
floatc	1.0	
floatd		Connected to the sequencer output. This will map to the ALPHA from the material_param's diffuse_reflectance anchor.

Table 5-31. *Float Sequencer Parameters*

maths → oscillators → float_sequencer		Second one
float_sequence	Covered in the following text	
length		Connect to the div module's output for the oscillator cycle duration
options		
behavior	trigger	
time_source	operating_system	
trigger		Connected to the pulse oscillator output
drive_type	time_internal_relative	

This sequencer will have a sequence as shown in Figure 5-17.

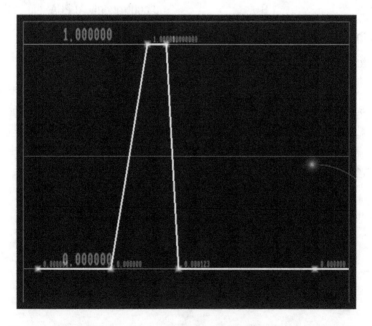

Figure 5-17. *Explosion, concentric ring visibility control, sequencer*

The anchors inside this control can be positioned manually by right-clicking them, respectively.

```
time = 0          value = 0

time = 0.249      value = 0
time = 0.372      value = 1.0
time = 0.436      value = 1.0
time = 0.476      value = 0.0
time = 1.0        value = 0.0
```

With all that setup done, the output will look like Figure 5-18.

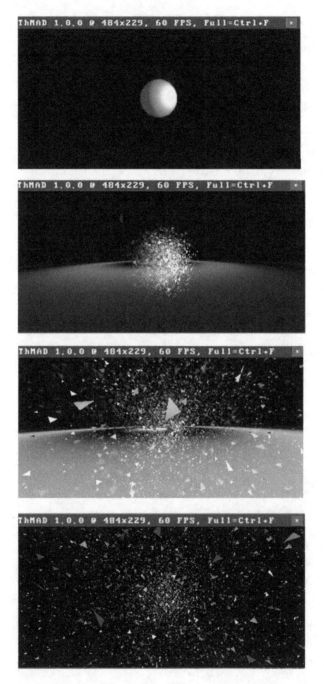

Figure 5-18. *Explosion output*

By the way, did you notice the concentric circles bend a little bit? This comes from the perspective view; in reality they are flat. You can change this bending phenomenon by playing around with the camera's parameters.

Explosions and Sound

The explosion from the previous section looks nice, but for obvious reasons it lacks the controllability by sound input. Sound is about the recurrence of beats, and explosions are usually singular events. But you can help out by multiplying the explosion and letting the exploding balls rotate.

In ThMAD there is nothing like a loop to multiply objects, but you can gather subpipelines in macros and then easily multiply them by cloning.

Note   The sample of this subsection is available under B-5.2_ Explosion_And_sound in the TheArtOfAudioVisualization folder.

To start, remove one of the interwoven planets from the previous section, and add a gl_translation module and a gl_rotation module. See Figure 5-19.

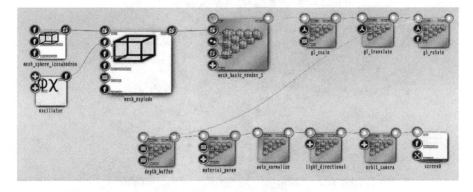

Figure 5-19. *Multi-explosion, macro preparation*

Decrease the size a little since you are going to have more objects in the end. Change the anchor scale of the gl_scale module to read 2.0; 2.0; 2.0.

You want to gather the complete top row of modules shown in Figure 5-19 into a macro. For that aim, you need to create an empty macro by right-clicking the canvas and selecting Create Macro. Open the macro for editing by right-clicking it and selecting Open/Close. Now select all the modules from the top row by drawing a rectangle around them. You can achieve that by pressing Ctrl, clicking in the top-left corner of a surrounding box, and dragging, while holding the mouse button to the bottom-right corner of the surrounding box. Release the mouse button.

To move the selected modules into the macro, press Shift+Ctrl, drag one of the previously selected modules, and release them over the macro. Your screen should now look like Figure 5-20.

Figure 5-20. *Multi-explosion, macro*

You later want to translate and rotate each macro instance individually, and you want to set the triggering frequency on a per- macro basis. To allow for that, you need incoming anchors for the macro. This is easy: just click each of the following anchors and drag it to an empty spot *inside* the macro space:

- Anchor freq of the oscillator module

- Anchor translation of the gl_translate module

- Anchor axis of the gl_rotate module

- Anchor angle of the gl_rotate module

The macro will now have four input anchors, as shown in Figure 5-21.

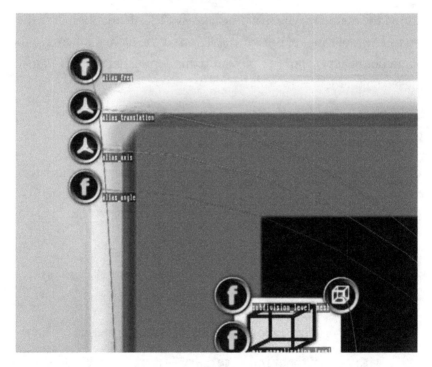

Figure 5-21. *Multi-explosion, macro with input anchors*

You can now collapse or close the macro since you do not need to see its interior any longer. Right-click it and then select Open/Close again.

Your state should now look like Figure 5-22.

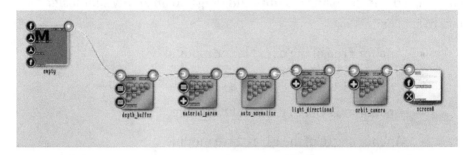

Figure 5-22. *Multi-explosion, state with macro*

Clone the macro maybe five times: press Ctrl+Alt and then drag the macro to an empty spot of the canvas. Connect the cloned macros to the input anchor of depth_buffer. Now add this connected module pair beside each macro:

- maths → arithmetics → binary → float → mod

- maths → arithmetics → binary → float → mult

Connect them and the output of mod with the alias_angle anchor of each adjacent macro. Set the param2 anchor of each mod module to 6.2830.

Add these modules:

- maths → arithmetics → binary → float → add

- maths → arithmetics → binary → float → mult

- system → time

- maths → accumulators → float_accumulator

- maths → arithmetics → ternary → float → mult_add

- sound → input_visualization_listener

Connect them as shown in Figure 5-23.

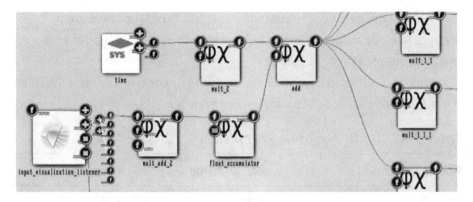

Figure 5-23. *Multi-explosion, angle control*

Add these modules:

- maths → converters → 4float_to_float4

- maths → arithmetics → binary → float → mult,
  three times

- maths → limiters → float_clamp

- maths → arithmetics → ternary → float → mult_add

Connect them as shown in Figure 5-24.

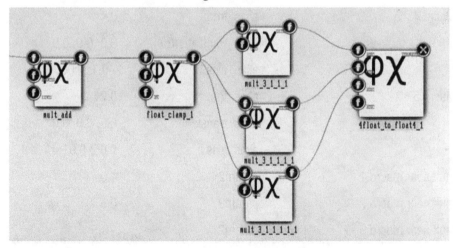

Figure 5-24. *Multi-explosion, color control*

Connect `4float_to_float4` to the `specular_reflectance` anchor of `material_param`, and connect `mult_add` to the `vu / vu_l` anchor of `input_visualization_listener`.

The possibilities of assigning values to the unconnected input anchors are endless; see Table 5-32 for a working example.

Table 5-32. *Multi-explosion Parameters*

Module Name	Anchor Name	Value
Macro 1	alias_freq	0.31
	alias_translation	10.0; 0.0; 0.0
	alias_axis	0.0; 0.0; 1.0
Macro 2	alias_freq	0.40
	alias_translation	−5.0; −3.0; 3.0
	alias_axis	0.01; 0.0; 1.0
Macro 3	alias_freq	0.27
	alias_translation	5.0; −6.0; 5.0
	alias_axis	0.02; 0.95; 0.30
Macro 4	alias_freq	0.4
	alias_translation	6.0; 0.0; 0.0
	alias_axis	−0.1; 0.66; 0.74
Macro 5	alias_freq	0.21
	alias_translation	0.0; 6.0; 0.0
	alias_axis	0.05; 0.91; −0.41
mult near macro 1	param2	1.0
mult near macro 2	param2	0.5
mult near macro 3	param2	−0.5

(*continued*)

Table 5-32. (*continued*)

Module Name	Anchor Name	Value
mult near macro 4	param2	−1.0
mult near macro 5	param2	−1.5
mult near time	param2	0.147
mult_add near float_accumulator	first_mult	0.048
	then_add	0.0
4float_to_float4	floatd	1.0
mult connected to floata of 4float_to_float4	param2	0.92
mult connected to floatb of 4float_to_float4	param2	0.92
mult connected to floatc of 4float_to_float4	param2	0.64
float_clamp	low	0.0
	high	1.0
mult_add near float_clamp	first_mult	0.3
	then_add	0.6

In the end, you will have endlessly exploding and re-appearing balls reacting to sound input.

Fractal Algorithms

Note   These samples are available under `B-5_3_Fractals_*` in the `TheArtOfAudioVisualization` folder.

Fractals are objects that live between the dimensions. They can be 1.3, 2.8, or even 0.1 dimensional. And they usually obey self-similarity, which means that patterns repeat endlessly if you zoom in or out.

In *Audio Visualization Using ThMAD*, you created a fractal via self-similarity; this time you will go the other way and use algorithms to create fractals. Fractals are sets of points; especially if you want to add some dynamics, the best candidates for fractal algorithms are particle systems. The following module takes an existing particle system and changes the particle coordinates according to an *iterated function system* (IFS) algorithm:

- `particlesystems → fractals → ifs_modifier`

The IFS algorithm takes a point p(x,y,z) and applies a function f to create a new point, $p \to f(p) = p'$. It does so all over, yielding $p'' = f(p')$, $p''' = f(p'')$, $p^{(4)} = f(p''')$, and so on, forever.

$$p = p(x,y,z) \to f(p) = p'$$

$$p' \to f(p) = p''$$

$$p'' \to f(p'') = p'''$$

...

Sound easy? Well, it is. The art is to define a good function f that behaves interestingly. The possibilities for functions bearing boring results or unstable functions that create runaways are endless. Fortunately, the `ifs_modifier` is accompanied by a set of working parameter sets you can use.

As of now, there is one restriction to the nature of the iteration function f inside ThMAD's `ifs_modifier` module: f takes its calculation rule from the function set of *affine transformations*. This means given a point p = p(x,y,z), f will do the following:

f chooses randomly with a 50% / 50% probability one of

$$x' = a_{11} \cdot x + a_{12} \cdot y + a_{13} \cdot z + at_1$$

$$x' = a_{21} \cdot x + a_{22} \cdot y + a_{23} \cdot z + at_2$$

$$x' = a_{31} \cdot x + a_{32} \cdot y + a_{33} \cdot z + at_3$$

or

$$x' = b_{11} \cdot x + b_{12} \cdot y + b_{13} \cdot z + bt_1$$

$$x' = b_{21} \cdot x + b_{22} \cdot y + b_{23} \cdot z + bt_2$$

$$x' = b_{31} \cdot x + b_{32} \cdot y + b_{33} \cdot z + bt_3$$

As an example, start with an empty canvas, which you can create by right-clicking and selecting New and then Empty Project. Place these modules and connect them as shown in Figure 5-25:

- renderers → opengl_modifiers → blend_mode (twice)

- renderers → opengl_modifiers → cameras → orbit_camera

- renderers → opengl_modifiers → light_ directional

- renderers → opengl_modifiers → depth_buffer

- renderers → opengl_modifiers → backface_ culling

- renderers → opengl_modifiers → material_param

- renderers → opengl_modifiers → gl_rotate_quat

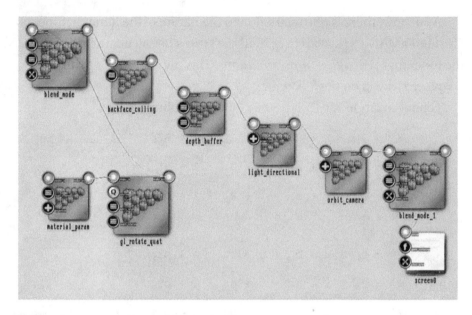

Figure 5-25. *IFS fractal, state basis*

For the parameters of these modules, see Tables 5-33 to 5-37.

Table 5-33. *Orbit Camera Parameters*

renderers → opengl_modifiers → cameras → orbit_camera		
rotation	0; 0; 1	
distance	23.0	You choose a larger distance and a small fov to avoid perspective artifacts.
fov	30.0	
perspective_correct	yes	The beams are white.

Table 5-34. *Light Directional Parameters*

renderers → opengl_modifiers → light_directional

enabled	YES
position	0; 0; 1
ambient_color	0; 0; 0; 1
diffuse_color	1; 0; 0; 1
specular_color	0.89; 0.81; 0.05; 1.0

Table 5-35. *Depth Buffer Parameters*

renderers → opengl_modifiers → depth_buffer

depth_test	DISABLED
depth_mask	DISABLED

Table 5-36. *Backface Culling Parameters*

renderers → opengl_modifiers → backface_culling

status	DISABLED

Table 5-37. *Material Param Parameters*

renderers → opengl_modifiers → material_param

ambient_reflectance	0.2; 0.2; 0.2; 1.0
diffuse_reflectance	0.11; 0.20; 0.26; 1.0
specular_reflectance	0.95; 0.87; 0.87; 1.0
emission_intensity	0; 0; 0; 1
specular_exponent	13.0

This is more or less the standard 3D rendering subpipeline. A characteristic of iterated functions systems is the point erratically jumping around and only its positions in the course of time building up the fractal. To avoid having the visualization appear too nervous, you can add an intense blurring effect. To do so, place the following modules on the canvas and connect them with each other and the already existing state modules as shown in Figure 5-26:

- renderers → basic → basic_textured_rectangle
- texture → effects → highblur
- texture → buffers → render_surface_single

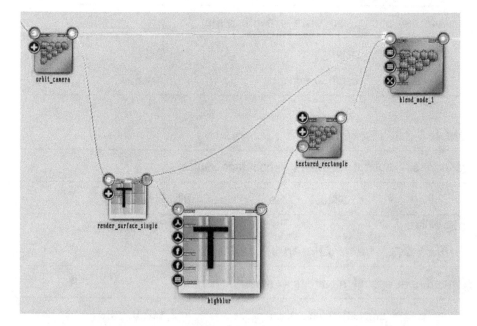

Figure 5-26. *IFS fractal, blurring modules*

Note the input order for the blend_mode module. The input from textured_rectangle must go *below* the input from the camera. To check this, double-click the input anchor. If necessary, you then could change the order by dragging one of the subanchors.

For the module parameter values, see Tables 5-38 to 5-40.

Table 5-38. *Textured Rectangle Parameters*

renderers → basic → textured_rectangle

color / global_alpha	0.97075	This is crucial for the blurring effect; the value must usually be slightly smaller than 1.0.
diffuse_reflectance	0.11; 0.20; 0.26; 1.0	
specular_reflectance	0.95; 0.87; 0.87; 1.0	
emission_intensity	0; 0; 0; 1	
specular_exponent	13.0	

Table 5-39. *Highblur Parameters*

texture_effects → highblur

translation	0.01; 0.005; 0	This is like a weak wind.
blowup_center	0.5; 0.5; 0	These are texture coordinates; hence, the 0.5, not the 0.0 of the object center.
blowup_rate	1000.0	The size of the radial blur effect.
attenuation	100.0	Controls the intensity of the blur effect.
texture_size	VIEWPORT_SIZE	This is important; the original image and blurred image must match in size and position.

Table 5-40. *Render Surface Single Parameters*

texture → buffers → render_surface_single

texture_size	VIEWPORT_SIZE	This is important; the original image and blurred image must match in size and position.

Next you can add the modules for the IFS object:

- renderers → particlesystems → simple

- texture → particles → blob

- particlesystems → generators → basic_spray_
  emitter

Connect them as shown in Figure 5-27.

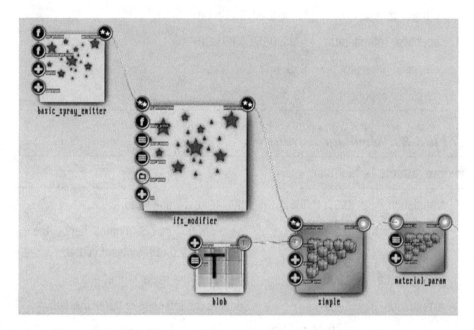

Figure 5-27. *IFS fractal, generator*

The parameters for the generator modules are shown in
Tables 5-41 to 5-43.

Table 5-41. *Blob Parameters*

texture → particle → blob

settings / alpha	yes	
settings / color	1; 1; 1; 1	Basic point color.
size	8x8	The particles are small here, so you do not need big textures.

Table 5-42. *IFS Modifier Parameters*

particlesystems → fractals → ifs_modifier

change_probability	1.0	If smaller than 1.0, not all particles will take part in the IFS-algorithm point coordinates will update with each frame. Since you are using blurring, you will always update.
change_random	off	If you select go instead, a random change of all IFS parameters gets triggered once. If the state gets saved, the value will always read off.
save_params	off	If you select go instead, the current IFS get saved inside the resources/ifs folder. If the state gets saved, the value will always read off.
load_params		Double-click this anchor to load an IFS parameter set from the resources folder. This is like a one-time trigger; the file name does not get persisted.
ifs / *		You can use this to freely set individual IFS parameters.

Table 5-43. *Basic Spray Emitter Parameters*

particlesystems → generators → basic_spray_emitter

num_particles	100000	The more the better. But be warned that with very high numbers, you can overload your computer.
spatial		
emitter_position	0; 0; 0	This is the root for the IFS algorithm. Changing these figures has no noticeable effect since the algorithm readily dictates the particles' positions.
speed / *	0.0 for all	Do not disturb the IFS algorithm with a particle speed.
size / particle_size_base	0.05	Increasing this might lead to overburdening your system; keep it small.
appearance		
color	Any	Has no influence.
time / particle_lifetime_ base	2.0	Has no big influence, but give the IFS some time to develop.
time / particle_lifetime_ random_weight	1.0	Lifetime randomization amount. It's important to avoid pumping effects (try setting this zero and restart to see what happens).

ThMAD IFSs live in three dimensions. To unleash the third dimension, you add a constant rotation around two axes. For that aim, add these modules:

- maths → arithmetics → binary → quaternion → quat_mul

- maths → arithmetics → functions → axis_ angle_to_quaternion (twice)

- maths → oscillators → oscillator

Connect them as shown in Figure 5-28.

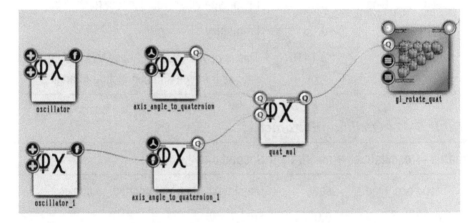

Figure 5-28. *IFS fractal, rotation*

For the parameters, see Tables 5-44 to 5-47.

Table 5-44. *Axis Angle to Quaternion Parameters*

maths → arithmetics → functions → axis_angle_to_quaternion		First one
axis	0; 1; 0	Just an example; choose at will
angle		Connected to an oscillator

Table 5-45. *Axis Angle to Quaternion Parameters*

maths → arithmetics → functions → axis_angle_to_quaternion		Second one
axis	0; 0; 1	Just an example; choose at will
angle		Connected to an oscillator

Table 5-46. *Oscillator Parameters*

maths → oscillators → oscillator		First one
osc / osc_type	saw	Usual type when you connect to angles.
osc / freq	0.05	Frequency.
osc / amp	3.1415	This is π; you usually choose this if using saw as the type and connecting to angles.

Table 5-47. *Oscillator Parameters*

maths → oscillators → oscillator		Second one
osc / osc_type	saw	Usual type when you connect to angles.
osc / freq	0.022	Frequency.
osc / amp	3.1415	This is π; you usually choose this if using saw as type and connecting to angles.

The output will look something like Figure 5-29.

Figure 5-29. *IFS fractal, output*

Note that the white saturation at the center region cannot easily be avoided since coordinate frequencies for IFSs usually obey logarithmic distribution laws and blending happens naturally in a linear distribution domain.

If you want to add sound responsiveness, the best place where you can hook into the `visualization_input_listener` module is the texture color of module `blob`.

Note   Take a look at the `B-5.3_Fractals_IFS_Sound` sample in the `TheArtOfAudioVisualization` folder for how this can be done.

Fire

To simulate candles, torches, flames, or fire in general, a kind of *plasma* is the conceptual tool of your choice. Physics has its own notion of plasma, which is rather hard to fully understand. The basis is a mutual interaction of negative and positive charged particles at different distance scales. For computer graphics, especially game development, different algorithms have been used to get a computationally inexpensive approximation of plasmas.

In ThMAD, a module named bitmaps → generators → subplasma uses an algorithm to mix an interpolation procedure between adjacent points and random numbers to generate plasma bitmaps. Later, you map this plasma onto a particle system, which mimics the gas movement in a flame.

The samples from this section are not described in a step-by-step manner. Instead, I will present an outline and ask you to go through the states provided in the installation directory.

Note   The states can be found in B-5.4_Fire_* in the TheArtOfAudioVisualization folder.

Start with the plasma as defined by state B-5_4_Fire_01PlasmaBitmap. It uses the bitmaps → generators → subplasma module to generate a plasma bitmap, as shown in Figure 5-30.

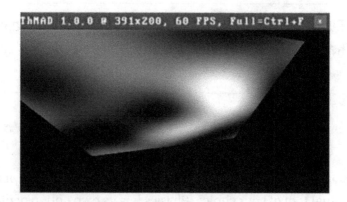

Figure 5-30. Plasma bitmap

You blend this plasma with a blob using the OVERLAY blend function. This is a blend mode that modulates the plasma, making the plasma brighter where the blob is bright and darker where the blob is dark. It uses the blob pixel brightness to decide whether to light or darken; this is why this blend mode is not symmetric. Order matters! The result is shown in Figure 5-31, and the name of the corresponding state is B-5.4_Fire_02PlasmaAndBlob.

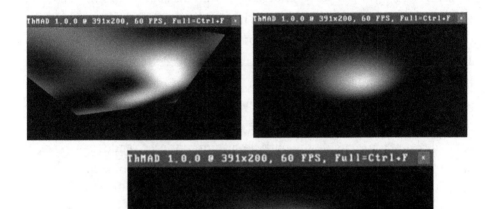

Figure 5-31. *Plasma and blob with OVERLAY blend function*

This basically centers the plasma at some point. This makes particles, covered next, align more smoothly.

The actual shape is random; you can play around with the rand_seed anchor of the subplasma module to try different shapes.

Next you allow for a rotation of the plasma blob. You will later use it as a texturing input for a particle system, so you first send the bitmap to the texture space of the graphics hardware and then do a rotation. Actually, texture rotation is always around the point with texture coordinates (0,0), but you want a rotation around the texture's center, which by definition is (0.5; 0.5). Remember, texture coordinates live in the square [0;0] → [1;1]. Thus, you translate (–0.5; –0.5), then rotate at some angle around (0;0), and finally shift back to (0.5; 0.5).

Fortunately, the current version of texture → modifiers → rotate does this all for you, so you do not need to add translation modules as was necessary in the predecessor of ThMAD. The corresponding state can be found in B-5.4_Fire_03TexRotate. What you do next is generate 100 random vertices via mesh → vertices → random_vertices, and use this and the texture from earlier as an input for particlesystems → generators → particles_mesh_spray. You add a translational movement to the particles via module particlesystems → modifiers → basic_ wind_deformer and inside the particle system renderer renderers → particlesystems → basic, you provide an elaborated color sequence. Figure 5-32 shows the output.

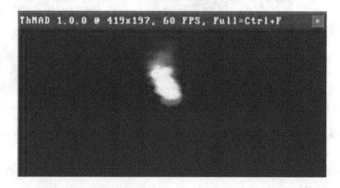

Figure 5-32. *Candle-like fire*

You can play with the values to change the appearance and intensity. Since there are so many parameters, a few hints follow so you do not get lost.

Changing the Overall Size

To change the overall size, go to the `random_vertices` module and change the contents of the `scaling` anchor. The x-coordinate there is interesting since it defines the horizontal base of the fire. A small value here will give you the impression of a candle, while a larger value will go in the direction of a bonfire. See Figure 5-33 (the left side is a small value for the x-coordinate of anchor `scaling` in module `random_vertices`, and the right has a bigger value).

Figure 5-33. *Changing the overall size of the fire base*

Changing the Intensity

The intensity can best be changed by altering the anchor `num_particles` of module `particles_mesh_spray`. See Figure 5-34 (left: 500 particles; right: 150 particles).

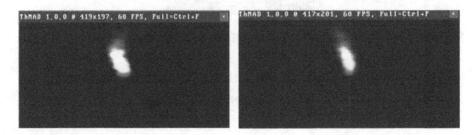

Figure 5-34. *Changing the intensity of the fire*

You can also change the particles' size at anchor spatial / size of module particles_mesh_spray.

Changing the Color Distribution

The color distribution inside the flame can be changed as well. It is a little trickier, though; you need to change the sequences inside the options anchor of module renderers → particlesystems → simple. The current set reads *_lifespan_ sequence (R, G, B, ALPHA); see Figure 5-35.

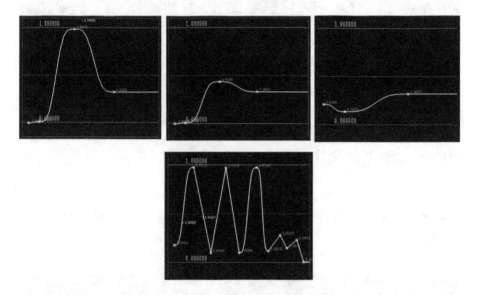

Figure 5-35. *Color distribution in flames, RGBA values*

You can see it starting at blue and then shifting to an orange and decaying to gray. The ALPHA value is somewhat erratic to improve the natural impression of the fire. The reason why it currently yields a bright white in the center of the flame and above lies in the blending mode set in module blend_mode; it is set to SRC_ALPHA / DEST_ALPHA, which will sum up to white if many objects overlap.

You can, for example, change the white to a more yellowish color if you change the blue sequence to something like shown in Figure 5-36.

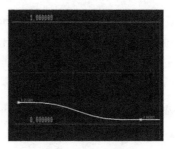

Figure 5-36. *Changing the blue channel*

The result will then look like Figure 5-37.

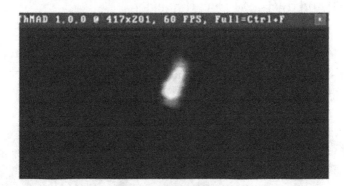

Figure 5-37. *Changing the colors of the fire*

The Problem of Sound Scaling

Sound comes in at varying levels. Maybe you change the PulseAudio sound server's level from time to time by altering the sound volume in the desktop's sound control. This can happen because you are also changing the amplification level of your stereo player and you want to compensate for that change. I do this quite often. The problem is that ThMAD cannot possibly know at what level the sound arrives at your ears just from

131

looking at PulseAudio's sound data that it receives. Or, which is another source for varying sound levels, different kinds of music enter your sound visualization.

In both cases, sound visualizations may break because when you build them, sound levels of one kind arrive at your visualization pipeline and then later different sound levels arrive.

A way to fix this is to collect sound levels in a moving average fashion and scale sound input by that. You'll now see an example of how this can be done.

Note   This and subsequent states of this section can be found at B-5.5_Sound_Level_Scaling* in the TheArtOfAudioVisualization folder.

Take a look at Figure 5-38, which shows the white curve of an incoming sound. This is the vu_l anchor of input_visualization_listener while some music is playing.

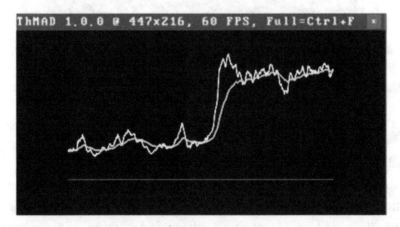

Figure 5-38. *Sound and smoothed sound*

The green line is the smoothed version, and it is obvious if you divide each white elongation by the corresponding smoothed green value that you would oscillate around 1.0, no matter if the overall volume is small as on the left side or high as on the other side. Doing this calculation, you will see an outcome like Figure 5-39.

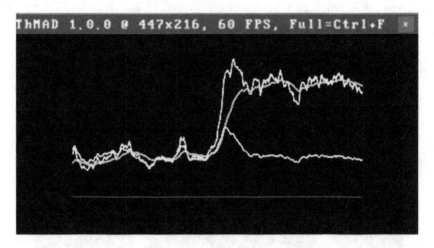

Figure 5-39. *Sound, smoothed sound and division*

Figure 5-40 shows the corresponding subpipeline. You can see that the yellow curve, coming from the dividing sound and smoothed sound, will still show the peaks while staying at the same overall level.

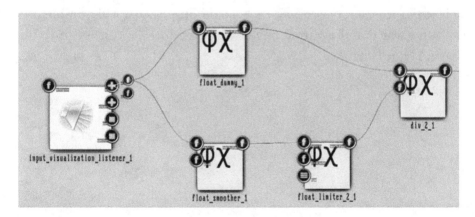

Figure 5-40. *Sound and smoothed sound, subpipeline*

There are two problems with that approach, though. The first is that you never want to divide by zero. That is why the float_limiter module was added; it is set to never let the value drop below 0.1. The other problem is that when the music starts to play, you will have a big peak, no matter how loud the music is; see Figure 5-41. The reason is that the smoothed value will build up slowly compared to the unsmoothed value, so the division will yield big numbers.

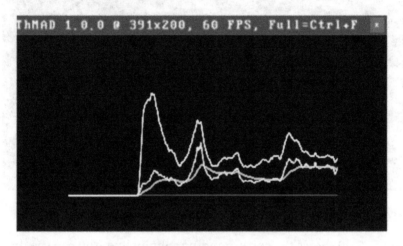

Figure 5-41. *Sound divided by smoothed sound, music onset*

If you want to avoid this, you can introduce the maths → arithmetics → unary → float → atan module, which will nonlinearly scale in a way that the output never will exceed 1.57 = $\pi/2$. See the blue line in Figure 5-42.

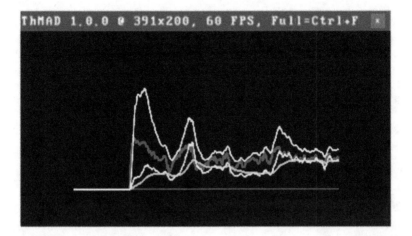

Figure 5-42. *Sound divided by smoothed sound, then atan scaler, music onset*

This atan function scaling smoothes big changes, which may be acceptable or even desirable or may be unwanted, depending on circumstances.

Another way to solve the scaling problem is to not look at the sound figures directly but instead at their derivatives in time. That is, you will look at the changes in time instead of at the absolute numbers. The corresponding sample state is named B-5.5_Sound_Level_Scaling_ Derivative, and a sample output is shown in Figure 5-43. The derivative is the green line.

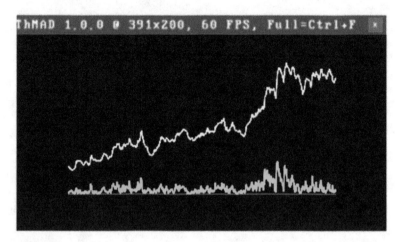

Figure 5-43. *Sound and sound derivative*

Note that in this case the float_smoother, mult, and abs modules had to be added besides the derivative module for a meaningful output.

A Space Odyssey

In the end sequence of the movie *2001: A Space Odyssey*, the viewer is led through a now famous psychedelic effect of two seemingly infinite planes with various shapes moving toward him. While the original effect's creation used up quite some budget and was technically challenging at the time the movie was made, you can try to use ThMAD to create something similar. It is not the intent here to copy the effect in detail, but instead you will stick to the main idea of two planes and shapes moving at high speed toward the viewer. You might even do better than the original.

Note   The sample of this section is available under B-5.6_A_ Space_Odyssey in the TheArtOfAudioVisualization folder. The shader code can be found at [ThMAD_INST]/share/thmad/ TheArtOfAudioVisualization-snippets/ B-5.6_A_Space_ Odyssey.

Two Planes

You start with positioning two planes of size 2×2 parallel to the x-y plane, centered at (0,0,+/– x) at a short distance of 2x. This is obviously not the same as using infinitely large planes, but if you choose a really small distance between the planes and position the camera very close to the edges, the illusion will be as if you had infinitely large planes.

These are the modules used for this example:

- renderers → basic → textured_rectangle (twice)
- renderers → opengl_modifiers → gl_translate (twice)
- renderers → opengl_modifiers → blend_mode
- renderers → opengl_modifiers → cameras → target_camera
- maths → converters → 3float_to_float3
- maths → dummies → float_dummy (renamed to camera_dist)

See Figure 5-44. Note that you are using float_dummy just for clarity and are renaming it to camera_dist since it gets connected to the y-position of the camera describing its distance to the edges of the plane.

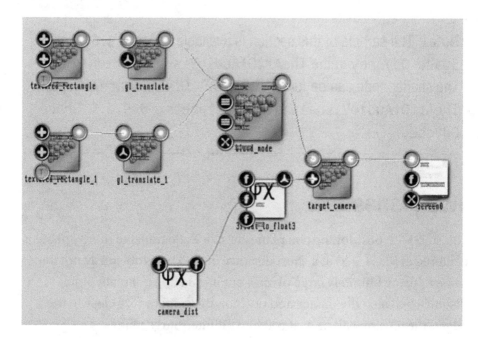

Figure 5-44. *A space odyssey: two planes*

The parameters are shown in Tables 5-48 to 5-54.

Table 5-48. *Textured Rectangle Parameters*

renderers → basic → textured_rectangle   Both of them
All parameters should be left at their default values.

Table 5-49. *Gl Translate Parameters*

renderers → opengl_modifiers → gl_translate		Upper plane
translation	0; 0; 0.02	Adding small values with the slider controls is not easy. It's best to enter the value directly at the knob in the second knob row. Then drag the slider to the lower end.

Table 5-50. *Gl Translate Parameters*

renderers → opengl_modifiers → gl_translate	Lower plane
translation	0; 0; −0.02

Table 5-51. *Blend Mode Parameters*

renderers → opengl_modifiers → blend_mode
All parameters should be left at their default values. The planes do not intersect.

Table 5-52. *Target Camera Parameters*

renderers → opengl_modifiers → cameras → target_camera		
camera/position		Connected to 3float_to_float3.
camera/destination	0; 0; 0	What you are looking at.
camera/upvector	0; 1.0; 0	Upright; the horizon is horizontal.
camera/fov	45.0	View angle. Make sure the horizon is wide enough.
camera/perspective_correct	no	You can try yes as well.

Table 5-53. *3Float To Float3 Parameters*

maths → converters → 3float_to_float3		Decomposing the camera position
floata	0.0	
floatb		You will be looking at the y=1 edges.
floatc	0.0	

Table 5-54. *Float Dummy Parameters*

maths → dummies → float_dummy		Renamed to camera_dist
float_in	1.00025	Specifies the y-position; see 3float_to_float3 and target_camera

As shown by the textured_rectangle modules you used, textures will hold the shapes that are going to be projected onto the planes.

Outgoing Texture Controllers

To have as much control as possible and also allow for dynamics, you provide for a texture control subpipeline. It consists of the following:

- texture → dummies → texture_dummy
- texture → modifiers → translate
- texture → modifiers → rotate
- maths → oscillators → oscillator (twice)

See Figure 5-45.

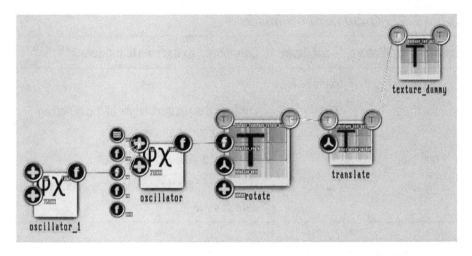

Figure 5-45. *A space odyssey, outgoing texture controls*

Note that you use oscillator_1 to control the amplitude of oscillator. The texture_dummy module is used just as an interface to the subsequent pipeline described previously. There you need two texture input channels and therefore here use the texture_dummy as a multiplexer.

The parameters are listed in Tables 5-55 to 5-58.

Table 5-55. *Translate Parameters*

texture → modifiers → translate

Leave at the default values. You can play around with these later.

Table 5-56. *Rotate Parameters*

texture → modifiers → rotate

rotation_angle	Connected to the first oscillator

Table 5-57. *Oscillator Parameters*

maths → oscillators → oscillator		Connected to the rotate module
osc/osc_type	triangle	
osc/freq	0.01	Controls the rotation angle of the outgoing texture orientation very slowly
osc/amp		Connected to the other oscillator
osc/ofs	0.0	Makes the straight view into the scene the default view

Table 5-58. *Oscillator Parameters*

maths → oscillators → oscillator		Connected to the other oscillator
osc/osc_type	triangle	
osc/freq	0.005	Slowly augments the amplitude of the other oscillator
osc/amp	0.25	Oscillates between 0 and 0.5
osc/ofs	0.25	

Connect the outgoing texture controller's subpipeline to the two planes' subpipeline from section "Two Planes" above by connecting the `texture_dummy` module with both `textured_rectangle` modules from earlier.

Texture Switcher

The main and most complicated module of the state is `selectors` → `texture_selector` to give some responsiveness to sound.

It comes together with these controllers:

- `maths → oscillators → oscillator, THREE TIMES`

- `maths → arithmetics → binary → float → mult`

- `maths → arithmetics → binary → float → div`

- `maths → interpolation → float_smoother`

- `sound → input_visualization_listener`

As the name of `texture_selector` says, it selects from several input textures. But it does a lot more; it allows for a smooth blending between adjacent textures in the incoming list, optionally wraps on index overflow or underflow, and even gives you the opportunity to define the shape of the blending function if blending is chosen as a mode.

They are connected as shown in Figure 5-46.

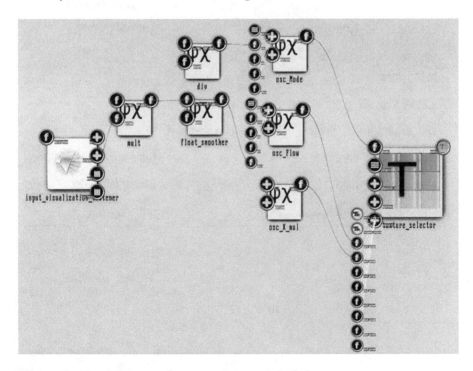

Figure 5-46. *A space odyssey, texture switcher*

The parameters and connections are shown in Table 5-59.

Table 5-59. *Texture Selector Parameters*

selectors → texture_selector		
index		The index of the texture to take. Connected to the osc_Mode oscillator.
inputs	8	You will later connect eight textures.
texture_x		Here the textures will get connected later.
options/Wrap	wrap	Wrapping mode on index underrun or overrun.
options/blend_type	Linear	Smooth blending.
options/blend_options/ blend_size	1024x1024	
shaders/vertex_program	See the following text	
shaders/fragment_program	See the following text	
shaders/shad_param 1		Connected to Osc_flow.
shaders/shad_param 2		Connected to Osc_X_mul.
shaders/shad_param 3..8		Unused.

The vertex_program anchor is as follows:

```
varying vec2 texcoord;
void main() {
  texcoord = gl_MultiTexCoord0.st;
  gl_Position = gl_ModelViewProjectionMatrix *
        gl_Vertex;
}
```

This is the default no-op vertex shader program; it just provides the texture coordinates to subsequent shaders and determines the standard vertex coordinates. The fragment_program anchor provides code for the fragment shader. The module determines the two neighboring textures given the index value and serves them to the fragment shader. Its code is as follows (the line numbers are for display purposes only):

```
(1) uniform sampler2D A_tex;
(2) uniform sampler2D B_tex;
(3) uniform float mode_index;
(4) uniform float A_mix; // weight of texture A
(5) uniform float B_mix; // weight of texture B
(6) uniform float shad_param1; // module params...
(7) uniform float shad_param2;
(8) uniform float shad_param3;
(9) uniform float shad_param4;
(10) uniform float shad_param5;
(11) uniform float shad_param6;
(12) uniform float shad_param7;
(13) uniform float shad_param8;
(14) varying vec2 texcoord; // from the vertex shader
(15) vec4 Acolorvec;
(16) vec4 Bcolorvec;
```

```
(17)
(18) void main(void) {
(19)    vec2 v = texcoord;
(20)    float f1 = v.y;
(21)
(22)    // acceleration factor along y
(23)    v.y = v.y * v.y * v.y * 3.0;
(24)
(25)    // y-flow
(26)    v.y = v.y - shad_param1;
(27)
(28)    // shrink along x
(29)    v.x = (v.x - 0.5) * shad_param2 + 0.5;
(30)
(31)    Acolorvec = texture2D(A_tex, v);
(32)    Acolorvec = vec4((Acolorvec[0] *
                A_mix), (Acolorvec[1] * A_mix),
                (Acolorvec[2] * A_mix),1.0);
(33) Bcolorvec = texture2D(B_tex, v);
(34) Bcolorvec = vec4((Bcolorvec[0] *
                B_mix), (Bcolorvec[1] * B_mix),
                (Bcolorvec[2] * B_mix),1.0);
(35) gl_FragColor = vec4(Acolorvec[0] +
                Bcolorvec[0],Acolorvec[1] +
                Bcolorvec[1],Acolorvec[2] +
                Bcolorvec[2],1.0);
(36) }
```

Lines 1 to 13 import values from the calling program, and line 14 imports the linearly mapped texture coordinates from the vertex shader. Lines 15 and 16 are just variable declarations. Starting line 22, you apply an acceleration to the y coordinates, which yields a somewhat physically unrealistic but nevertheless impressing acceleration of shapes flying toward the observer. Line 26 does not look that impressive, but in fact it creates the main effect of flowing along the y-axis. The flow speed is controlled by import parameter shad_param2. Line 28 fetches the module import parameter shad_par1 and from that scales along the x-axis. Everything starting at line 31 performs the blending by combining color values from the two textures provided by the module for the current blending and finally sets the pixel color.

Because of this shader code and linear being chosen as a blending mode, the desired smooth blending will happen. The idea is as follows: if the index is, for example, 3.4, this means the texture number 3 will be assigned the weight 0.4 = 40%, the texture number 4 will be assigned the weight 1 – 0.4 = 0.6 = 60%, and the blend will be 0.4 * texture3 + 0.6 * texture4.

As soon as you are finished with this subpipeline, connect the output from the texture_selector module to the input of the rotate module of the outgoing texture control subpipeline in section "Outgoing Texture Controllers" above.

Shape Creation Textures

For the input of the texture_selector module, you will use eight subpipelines. Since they are pretty straightforward and do nothing special apart from presenting image data in a texture, this will be only a concise summary of their characteristics.

Note   For details, please take a look at B-5.6_A_Space_Odyssey in the TheArtOfAudioVisualization folder.

Of course, feel free to create your own shapes. The only requirement is that any module creates a texture of size 1024×1024. The texture generators used for my version of the state are as follows:

- A simple white blob created using `bitmap → generators → blob`. Make sure its `alpha` anchor is set to `no`. It is converted to a texture via `texture → loaders → bitmap2texture`.

- A red star using the same modules as above but with different parameters

- A cloudy scene using `bitmap → generators → perlin_noise` and then again `texture → loaders → bitmap2texture`.

- A `texture → loaders → png_tex_load` module for directly presenting a PNG image file to the texture image data. You could load any PNG file, but usually it is better to have it sized 128×128, 256×256, 512×512, or 1024×1024. I loaded a precomputed image with a randomly distributed set of triangles.

- A set of concentric circles from `texture → particles → concentric_circles`.

- A blue rectangle from `renderers → basic → colored_rectangle`. The renderers data is loaded to a texture buffer via the module `texture → buffers → render_surface_color_buffer`. Note that because of the way ThMAD works internally, the colored rectangle gets painted only once and then updated to the texture buffer. If you want to play around with the rectangle, change its size, color, angle, or whatever, in order not to fill up the texture buffer with garbage. Another black rectangle filling the complete space should be preceded.

- The same as the concentric circles but several randomly distributed yellow rectangles with ALPHA < 1 and a green background.

- A mesh → particles → mesh_rays with adjacent renderers → mesh → mesh_basic_render creating some ray-like blue structure.

Connect all these textures to the items of the texture_x subanchor of the texture_selector module in section "Texture Switcher" above.

All Combined

Because of the oscillator-controlled rotation *inside* the texture, you can do better than the original that has only one view. The scene feels like you are flying between two planes with hypersonic speed. All the dynamics happen inside the graphics hardware; the textures get uploaded only once when the state starts its work. This allows for the impressive speed, but please be warned that it nevertheless goes to the edge of what a modern onboard graphics controller can handle. If the video is too unstable for you, maybe try to remove some of the textures or lower the speed; remember, it is the shad_param1 anchor value of module texture_selector.

Figure 5-47 shows a bird's-view of the complete state and a snapshot of the output in Figure 5-48.

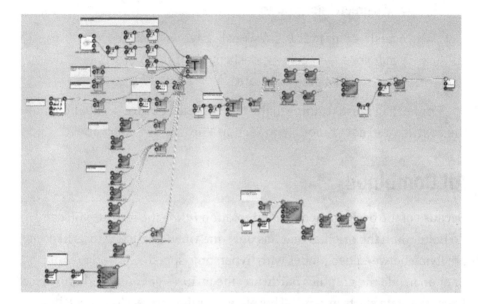

Figure 5-47. *A space odyssey, complete state*

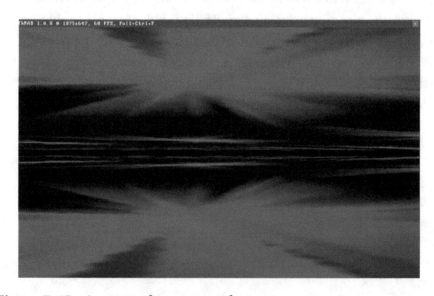

Figure 5-48. *A space odyssey, snapshot*

Of course, you should see it in action or at least watch the video on YouTube; search for *PMSSpaceOdyssey2017*.

Making Sequences

ThMAD currently has no elaborate sequencing functionalities as you'd expect from a video or game engine. You learned that the player switches between visuals, but this happens on a random basis and only scarcely counts as sequencing.

Fortunately, ThMAD has a few modules that, if used the right way, might serve some basic sequencing needs. More precisely, it is possible to use the subpipeline blocker modules, which can be controlled by input anchors.

Here you will see an example that switches from a system with a particle system with bigger textures, running on random point sources, to a system with a particle system running with smaller particles originating from a single source point. See Figure 5-49.

Figure 5-49. *Sequence control, base image, and add-on sequences*

Particle systems were introduced in Chapter 2, so this section just refers to the states given in the installation and points out some important issues.

Note   Examples from this section are available under B-5.7_ Sequencing* in the TheArtOfAudioVisualization folder.

The states B-5.7_Sequencing_A, B-5.7_Sequencing_B, and B-5.7_ Sequencing_C contain static images from a photo and the first and second particle systems, respectively. The state B-5.7_Sequencing contains the sequencing. There, a saw oscillator produces the output. Two instances of the module maths → arithmetics → unary → float → ifinside take the oscillator's output and produce a 1.0 if it is inside [0;0.5] or [0.5;1], respectively. That means the first ifinside produces a 1.1 if the oscillator is between 0 and 0.5, and the second ifinside produces a 1.0 if the oscillator is between 0.5 and 1.0. The rest of the time each of the modules produces 0.0. If you interpret 1.0 as on and 0.0 as off, the branches get mutually switched on and off. The actual control is performed by two modules of type system → blocker that enable a subpipeline if the input is greater than 0.5 and otherwise disable it. Figure 5-50 shows the controlling part of the state.

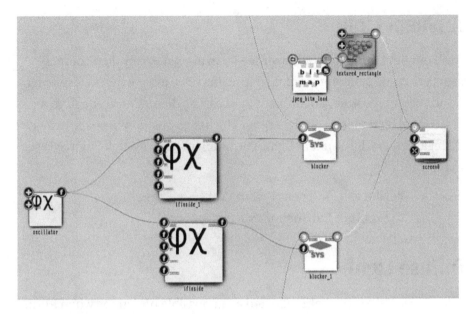

Figure 5-50. *Sequence control*

Of course, you can make extended sequences if you add more ifinside modules and divide [0;1] into smaller chunks.

Lighting Revisited

In *Audio Visualization Using ThMAD*, I focused on the theoretical aspects of lighting, but I reckon it is worthwhile to revisit that topic from a practical point of view.

Note The examples from this section are available under B-5.8_* in the TheArtOfAudioVisualization folder.

The different light sources are described in the following sections.

Ambient Light

Ambient light is the dull light you see in an almost totally dark room. It seems to come from nowhere, and it seems to be everywhere. Despite its nondirectional nature, an ambient light can be defined to come from a directional light placed on the scene. The light's position in space just gets disregarded for the ambient light. The ambient part of a light source is set in this module:

- renderers → opengl_modifiers → light_ directional at anchor ambient_color

Diffuse Light

The diffuse light is the light reflected at all possible directions when hitting a surface. It gets set at the same place as the ambient light, in this module:

- renderers → opengl_modifiers → light_ directional at anchor diffuse_color

Specular Light

Specular light gets reflected more or less sharply at a defined angle when light beams hit a surface. It gets defined here:

- module renderers → opengl_modifiers → light_ directional at anchor specular_color

Clear Color

Another source of color or light is the screen's clear color. It will be applied to the whole scene in each frame before anything else gets rendered. It will thus show up as a background color and may shimmer through objects only if they are in part transparent. The screen's clear color gets set in the screen0 module at anchor clear_color.

Material Ambient Color

The ambient light defined by a light source has its counterpart in an ambient reflection color of a surface material. There it can be configured directly using this module:

- renderers → opengl_modifiers → material_param at anchor ambient_reflectance

The ambient part of the outcome color will be the product of each of the RGB components from the ambient color coming from the light and the ambient color defined as a material parameter. An ALPHA value plays a role there as well, but it cannot be defined as an ambient material parameter; instead, the ALPHA value for the diffuse reflection material parameter will be used for the ambient color calculation as well. See Figure 5-51, which shows a cube with ambient light; it has a material parameter ambient light RGB composition of YELLOW = RED + GREEN = (1, 1, 0) and a directional light ambient color of GREEN = (0, 1, 0). You get a resulting ambient color after the part-wise multiplication (1, 1, 0) · (0, 1, 0) = (0, 1, 0), which is a plain GREEN.

Caution   As for the influence of the material parameters' ambient RGB values, each color component will be multiplied with the ALPHA value of the diffuse light component configured for the material.

Note that for realistic scenes an ambient influence of 10 percent maximum will do in most cases. Since this cannot be adjusted by using the ambient material parameter ALPHA, you will have to control this with either low RGB values inside the material parameter module's settings or with low RGB values inside the light's ambient color settings.

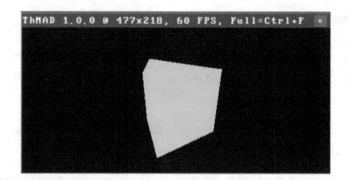

Figure 5-51. *Ambient light*

Material Diffuse Color

Another material parameter is the diffuse reflectance parameter, defined in this module:

- renderers → opengl_modifiers → material_param
  at anchor diffuse_reflectance

The diffuse light outcome of some material point is calculated by multiplying each color component of the incoming diffuse light with the matching material parameter's diffuse reflectance color component. While for the ambient lighting the position of lights and surfaces in space does not play a role, for diffuse light the 3D nature of objects prevails. Figure 5-52 shows a cube with diffuse light. It has a material parameter diffuse light RGB composition of YELLOW = RED + GREEN = (1, 1, 0) and a directional light diffuse color of GREEN = (0, 1, 0). You get a resulting diffuse color after the part-wise multiplication (1, 1, 0) · (0, 1, 0) = (0, 1, 0), which is a plain GREEN.

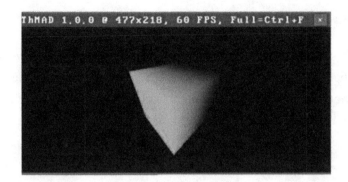

Figure 5-52. *Diffuse light*

Note that for this to work correctly, any surface element needs to have a normal vector defined. ThMAD does its best to provide the normal vectors needed, but if they are missing, unexpected phenomena occur with light shining on the surface element. On the other hand, normal vectors can be messed with, and interesting effects may be accomplished that way. See, for example, Figure 5-53.

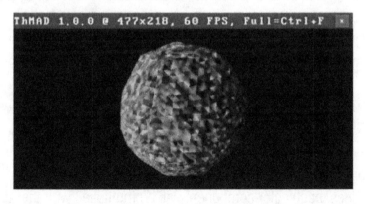

Figure 5-53. *A ball with diffuse lighting and random normal distortion*

Material Specular Color

While diffuse light reflects in all directions when hitting a surface point, specular light gets reflected, like with a mirror, at precisely the same angle on the other side. The calculation is the same as for ambient or diffuse light. Each material parameter's RGB component at the following location gets multiplied with the corresponding RGB component of the light module:

- renderers → opengl_modifiers → material_param at anchor specular_reflectance

In Figure 5-54, because the specular light is white (1,1,1) for both material parameters and light module, the specular part gets added as white to the green component from the diffuse light calculation.

Especially for specular reflectance, you additionally have the concept of shininess, which describes the amount of fuzziness regarding the viewer's position and the reflected light beam. It is defined by the specular_exponent anchor of module material_param and can range from 0 to 128. For 0 you have the minimum shininess with the least angular match dependency, and for 128 you have the maximum shininess with a maximum angular match dependency, as shown in Figure 5-55. The left ball has a specular exponent of 5, and the right ball has a specular exponent of 50.

Caution   As for the influence of the material parameters' specular RGB values, each color component will be multiplied with the ALPHA value of the diffuse light component configured for the material.

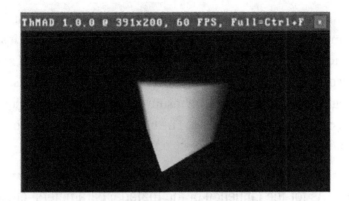

Figure 5-54. *Specular light*

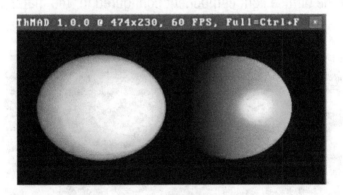

Figure 5-55. *Specular light exponent, or shininess*

Emissive Light

Materials can produce light without the help of a dedicated light switched on. Still, in ThMAD you need to introduce the lighting module for emissive light to work. And it must be enabled, but all the other parameters may be set to zero. The emissive light itself is defined in this module:

- renderers → opengl_modifiers → material_param at anchor emission_intensity

The emissive light is similar to the ambient light contribution but does not get combined with a corresponding light source parameter. It stands for itself and adds to all the other light produced by the material and light source. Also, it does not produce any reflections on other objects. For emissive light, see Figure 5-56. A light source is added and enabled, but it does not contribute to the output. The emissive light comes from the material parameters alone.

Caution   As for the influence of the material parameters' emissive RGB values, each color component will be multiplied with the ALPHA value of the diffuse light component configured for the material.

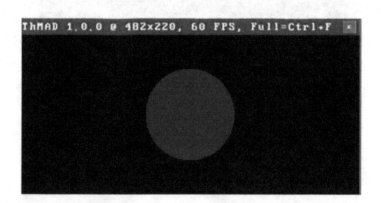

Figure 5-56. *Emissive light*

Something that has not yet been mentioned is the lighting model. In some cases, the standard model can be used, which means you do not have to do anything with respect to the lighting model. As an example for altering this mode, consider two balls of radius 0.5 placed at (−0.6;0;0) and (+0.6;0;0), with a specular light at (0;0;1) and a camera at (0;0;2). With the usual 3D setup, you will end up with something like shown in Figure 5-57.

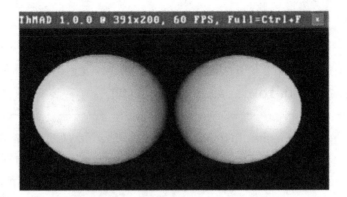

Figure 5-57. *Specular lighting, standard color model*

While from a perspective point of view this is correct, you might under certain circumstances want to have the specular spots in the projected middle, maybe like simulating glowing pupils looking straight at you. This can be achieved by using the so-called Local Viewer light model, and the module that allows you to use it is the following:

- renderers → opengl_modifiers → light_model

Now if you introduce this module into the state right before the light_directional module and set its anchor local_viewer to EYE_COORDS, the light spots end up like you want, as shown in Figure 5-58.

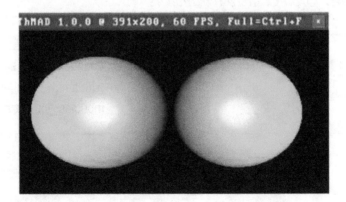

Figure 5-58. *Specular lighting, Local Viewer model*

161

Ocean Revisited

Note   This example is available under B-5.9_Ocean_And_Sky in
the TheArtOfAudioVisualization folder.

With the ocean example from *Audio Visualization Using ThMAD* and
what you learned about Perlin noise there, you can improve that ocean
view by adding some clouds to the sky. You start with the ocean module
system from that book but change a couple of parameters so that they now
match Tables 5-60 to 5-67.

Table 5-60. *Screen0 Parameters*

screen0		
clear_color	0.83; 0.92; 1.0; 1.0	It is actually easier to let the clouds be represented by the background color and later add Perlin noise to represent "negative" clouds. This is why the screen clear color is set to almost white now.

Table 5-61. *Orbit Camera Parameters*

renderers → opengl_modifiers → cameras → orbit_camera		
rotation	0.0; −1.0; 0.03	Changed a little, so together with a new translation modules, you get more sky
distance	20.0	
fov	50.7	

Table 5-62. *Light Directional Parameters*

renderers → opengl_modifiers → light_directional		Unchanged from the original ocean state
position	−0.19; 0.71; 1.0	
ambient_color	0.16; 0.22; 0.23; 0.42	Turquoise ambient color
diffuse_color	0.94; 0.92; 1.0; 1.0	Ocean blue diffuse color
specular_color	0.95; 0.94; 0.81; 1.0	A little yellowish bright specular

Table 5-63. *Material Param Parameters*

renderers → opengl_modifiers → material_param		Unchanged from the original ocean state
ambient_reflectance	0.21; 0.19; 0.06; 0.98	Dark yellow ambient Reflectance
diffuse_reflectance	0.03; 0.13; 0.83; 1.0	Blue diffuse reflectance
specular_reflectance	0.92; 0.91; 0.68; 1.0	Light yellow specular reflectance
specular_exponent	20.0	

Table 5-64. *Backface Culling Parameters*

renderers → opengl_modifiers → backface_culling		Unchanged from the original ocean state
status	DISABLED	Needs to be disabled because of an internal bug

Table 5-65. *Depth Buffer Parameters*

renderers → opengl_modifiers → depth_buffer		Unchanged from the original ocean state
depth_test	ENABLED	
depth_mask	ENABLED	

Table 5-66. *Mesh Basic Render Parameters*

renderers → mesh → mesh_basic_render		Unchanged from the original ocean state
vertex_colors	no	Let the material_param module define the color
use_display_list	no	
use_vertex_colors	no	
particles_size_center	no	
particles_size_from_color	no	
ignore_uvs_in_vbo_updates	no	

Table 5-67. *Ocean Parameters*

mesh → generators ocean		Unchanged from the original ocean state
time_speed	3.2	Time multiplier for the wave movement
wind_speed_x	5.0	
wind_speed_y	12.0	
lambda	1.12	Wave speed
normals_only	no	

Note   You can find under `A-5.1.3_Ocean` in the
`TheArtOfAudioVisualization` folder.

What you now do is insert these modules between the `mesh_basic_`
`render` and the `depth_buffer`:

- `renderers → opengl_modifiers → translate`

- `renderers → opengl_modifiers → blend_mode`

See Figure 5-59.

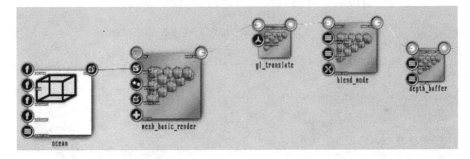

Figure 5-59. *Enhanced ocean view, tanslation and blending added*

The translation module is for lowering the horizon a little since the sky
gets more interesting now. The blending module is for later combining the
ocean and sky. The parameters are listed in Tables 5-68 to 5-69.

Table 5-68. *Gl Translate Parameters*

renderers → opengl_modifiers → gl_translate		
translation	−2.8; 5.0; −2.5	Lower the horizon

Table 5-69. *Blend Mode Parameters*

renderers → opengl_modifiers → blend_mode
Leave at the default values

Now you create the sky subpipeline. Place the following on the canvas, and connect them all as shown in Figure 5-60:

- `bitmaps → generators → perlin_noise`
- `texture → loaders → bitmap2texture`
- `renderers → basic → textured_rectangle`
- `renderers → opengl_modifiers → gl_rotate`
- `renderers → opengl_modifiers → gl_scale`

Figure 5-60. *Enhanced ocean view, sky subpipeline*

The intention here is to project Perlin noise onto a sheet covering the sky in an upright position at the end of the ocean.

The parameters are as shown in Tables 5-70 to 5-73.

Table 5-70. *Gl Scale Parameters*

renderers → opengl_modifiers → gl_scale		
scale	2.6; 1.9; 0.9	Make big enough to cover the whole sky

Table 5-71. *Gl Rotate Parameters*

renderers → opengl_modifiers → gl_rotate		
axis	13.3; 0.06; 0.03	Rotate to stand upright
angle	0.76	

Table 5-72. *Textured Rectangle Parameters*

renderers → basic → textured_rectangle		**The projection surface for the Perlin noise**
position	0.0; −4.7; 0.0	Rotate to stand upright
size	10.5	
angle	0.0	

Table 5-73. *Perlin Noise Parameters*

bitmaps → generators → perlin_noise		**Generates the clouds**
perlin_options		
rand_seed	4.0	Changes to alter cloud shapes
perlin_strength	1.8	Changes to alter cloud intensity
size	256x256	
octave	6	Detail level
frequency	4	Noise scale
color	1.0; 1.0; 1.0; 1.0	Negative cloud color
alpha	yes	Makes transparent to let open sky shine through

As a last step, connect the output from gl_scale to the input from blend_mode, such that blend_mode now has two inputs. Order matters; to make sure it is correct, double-click the input anchor and let the sky be the second one, as shown in Figure 5-61 (the sky subpipeline connects from below, which is not shown here).

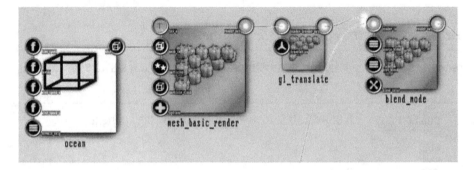

Figure 5-61. *Enhanced ocean view, blend_mode order*

The output will look like Figure 5-62. As usual, you can switch to full-window mode by pressing Ctrl+F (also hiding the info header there by pressing Alt+T).

Figure 5-62. *Enhanced ocean view*

Summary

In this chapter, you investigated a couple of states using advanced concepts and extending stories from *Audio Visualization Using ThMAD*. You looked at textures, created a state with exploding objects, saw a fractal algorithm, learned about fire and sound scaling issues, built a state with sequences, looked more intensely at lighting issues, and improved the ocean view from *Audio Visualization Using ThMAD*.

In the next chapter, you look at advanced states using shader constructs.

CHAPTER 6

Stories II

The stories in this chapter are examples of using only a few modules; instead, a lot of functionality is in the shader code. See Chapter 4 for an introduction to shaders.

Color Gradient Mapping

You can take an interesting picture uploaded as a texture to the graphics hardware, convert its colors to grayscale values, and map those to a color gradient that gets calculated in the shader. This mapping gets controlled from the outside using sound and uniform variables.

Since the vertex shader doesn't know anything about pixel-wise colors, you will provide just the no-op vertex shader, move the effect onto the fragment shader, and otherwise map the texture onto a rectangle.

In a multitexture environment, you'd probably prefer providing the complete color gradient inside another texture, but ThMAD 1.1 doesn't yet know how to handle multiple textures, so you will move the color gradient calculation to the shader, hoping it will not take away too much computing power.

© Peter Späth 2018
P. Späth, *Advanced Audio Visualization Using ThMAD*,
https://doi.org/10.1007/978-1-4842-3504-1_6

Color Gradient Algorithm

You first pick an interesting color gradient as a set of RGB values. Instead of just guessing a color gradient or taking one of the preexisting color gradients on the Web, you can take a more artistic approach. You can pick the most prominent colors from an image, convert them into the HSV color space, use a Monte Carlo algorithm to sort the colors, and write shader code based on that.

So, you start with a picture; Figure 6-1 shows the one I chose.

Figure 6-1. *Color gradient extraction*

This image gets loaded by a Groovy script, the color information gets extracted, and the Monte Carlo sorting algorithm is applied.

Note   This script is available under /opt/thmad/share/thmad/
TheArtOfAudioVisualization-snippets/B-6.1_Color_
Gradient_Mapping/gradient.groovy.

```
import javax.imageio.* import java.awt.image.*
import java.util.concurrent.atomic.AtomicInteger import java.
awt.*

IMAGE = "/home/peter/Desktop/001.png" // use your own
SUBDIVIDE = 50

STARTPOINTS = 100
TEMPSTART = 10.0
TEMPSTEP = 0.999995
MAX_NO_ACCEPT_COUNT = 1000

BufferedImage img = ImageIO.read(new File(IMAGE)) WIDTH = img.
width
HEIGHT = img.height

// -------------------------------------------------
// From HSV values from [0;1]^3, subdivide into
// buckets of size        1/SUBDIVIDE * 1/SUBDIVIDE *
// 1/SUBDIVIDE. With now each HSV component from
// [0,1,2,3,...,SUBDIVIDE-1] we can build a single
// integer specifying the HSV coordinates
def hsvToCoord(def f3) {
    (int)(0.999 * f3[0] * SUBDIVIDE) *
        SUBDIVIDE*SUBDIVIDE +
    (int)(0.999 * f3[1] * SUBDIVIDE) * SUBDIVIDE
    + (int)(0.999 * f3[2] * SUBDIVIDE) }
```

```
// --------------------------------------------------------
// Coord back to HSV
def coordToHsv(def coord) {
    def cc = coord
    def h = (int)(cc / SUBDIVIDE / SUBDIVIDE) *
            1.0/SUBDIVIDE
    cc -= h * SUBDIVIDE * SUBDIVIDE * SUBDIVIDE
    def s = (int)(cc / SUBDIVIDE) * 1.0 / SUBDIVIDE
    cc -= s * SUBDIVIDE * SUBDIVIDE
    def v = (int)(cc) * 1.0 / SUBDIVIDE
    [h, s, v]
}

// --------------------------------------------------------
// A distance in HSV space (Hue is rolling)
def hsvDist(def hsv1, def hsv2) {
    def h1 = Math.min( hsv1[0], hsv2[0] )
    def h2 = Math.max( hsv1[0], hsv2[0] )
    def hDist = Math.min( h2 - h1, 1.0 - h2 + h1 )
    def sDist = Math.abs(hsv1[1] -
    hsv2[1]) def vDist = Math.abs(hsv1[2]
    - hsv2[2]) Math.sqrt( hDist*hDist +
    sDist*sDist +
              vDist*vDist )
}

// --------------------------------------------------------
// Collect histogram. This is a mapping HSV-Coord →
// number
Map histo = [:]
```

```
for( int i = 0; i < WIDTH; i++ ) {
    for( int j = 0; j < HEIGHT; j++ ){
        int pix = img.getRGB( i, j )
        int a = (pix & 0xFF000000) >> 24
        int r = (pix & 0xFF0000) >> 16
        int g = (pix & 0xFF00) >> 8
        int b = pix & 0xFF

        float[] hsv = new float[3]
        Color.RGBtoHSB(r,g,b,hsv)
        int c = hsvToCoord(hsv)
        if(!histo[c]) histo[c] = new AtomicInteger(0)
            histo[c].incrementAndGet()
    } }

// -----------------------------------------------------
// Collect 100 most important HSV buckets. The
// inject() creates a list of [Number;HSV-Coord]
// pairs, the sort() sorts according to the
// frequency, the take() takes only the head part of
// the list, and the collect() maps to HSV-triples
def hsvPoints = histo.entrySet().inject([],
{ bas, inj ->
            int num = inj.value
            int hsv = inj.key
            bas.add([num, hsv])
            bas
    }).sort { -it[0] }.take(STARTPOINTS).
collect{ coordToHsv(it[1]) }
```

```
// -------------------------------------------------------
// Find HSV points the most apart def maxDist = -1.0
def p1 = 0
def p2 = 0
for(int i = 0; i < hsvPoints.size(); i++) {
  for(int j = i+1; j < hsvPoints.size(); j++) {
      def d = hsvDist(hsvPoints[i],
             hsvPoints[j]) if( d > maxDist) {
                  p1 = i
                  p2 = j

                  maxDist = d
              }
          }
}

// ------------------------------------------------------
// Build work array for Monte Carlo algorithm. First
// and last fixed
def arr = [p1]
for(int i = 0; i < hsvPoints.size(); i++)
      if(i != p2 && i != p1)
arr.add(i) arr.add(p2)
arr = arr.collect { hsvPoints[it] }

// ------------------------------------------------------
// Calculate energy = total way in HSV space
def calcEner(def arr1)
      { def e = 0.0
      for(int i = 0; i < arr1.size() - 1; i++)
            e += hsvDist(arr1[i], arr1[i+1])
      e
}
```

```
// -------------------------------------------------
// A test step, just swap random indices
def step(def arr) {
    int i1 = 1 + Math.random() * (arr.size() -
    2) int i2 = 1 + Math.random() * (arr.size()
    - 2) def tmp = arr[i1]
    arr[i1] = arr[i2]
    arr[i2] = tmp [i1, i2]
}

// -------------------------------------------------
// Revert step
def unstep(def arr, def swapped)
    { def tmp = arr[swapped[0]]
    arr[swapped[0]] = arr[swapped[1]]
    arr[swapped[1]] = tmp
}

// -------------------------------------------------
// Monte Carlo algorithm, simulated annealing
double temp = TEMPSTART
int noAcceptCount = 0

int cnt = 0
def energy = Double.MAX_VALUE
while(noAcceptCount < MAX_NO_ACCEPT_COUNT) {
    cnt++
    def swapped = step(arr)
    def newEner = calcEner(arr)
```

```
        def accept = false
        if(newEner < energy) {
                accept = true
        } else if(newEner != energy) {
                def tst =
                        Math.exp( -(newEner-energy) /
                temp) if(Math.random() < tst)
                        accept = true
        }

        if(accept) {
                energy = newEner
                noAcceptCount = 0
                if((cnt%1000) == 0)
                        println("${cnt} ${temp} ${energy}")
        } else {
                unstep(arr, swapped)
                noAcceptCount++
        }

        temp *= TEMPSTEP
}
// --------------------------------------------------
// DONE
println(arr.collect{it.join(",")}.join(","))
```

In this script, the first parts loads the image from a PNG file. Then you collect all the pixels from the image, convert them to the HSV color space, and count the occurrences in grid buckets of equal sizes inside the HSV space. You thus get a histogram of occurrences mapped to HSV bucket coordinates. You take only the 100 most important buckets, thus favoring the most prominent colors from the image. Now you need some kind of sorting, and for this you apply a Monte Carlo simulated annealing algorithm. Without going into too much detail here, the algorithm consists of a minimization of the distance to walk inside the HSV space to get from the first point to the last point, allowing for swapping random pairs to find that minimum. This net distance is called *energy* inside the script. Once in a while, however, a temporarily higher energy is allowed, and this gets controlled by the *temperature* parameter. The higher the temperature, the higher the probability of accepting temporarily higher energies. The temperature decreases steadily while the algorithm is at work. In the end, you will have a minimized energy and an HSV color palette with adjacent colors being similar. The output is a list of HSV values that will be used later in the shader code. The corresponding palette might look like Figure 6-2.

Figure 6-2. *Calculated color gradient*

Note   The possibilities for improving the algorithm are endless, but so is the time you will need to find the best code. You might try changing the step() function to allow for more elaborate steps compared to just swapping two random colors. Also, you can apply smoothing algorithms at the end.

To see how to run Groovy code, go to the Groovy home page at www. groovy-lang.org/.

You will now continue with the state you need to create. Place the following modules on the canvas and connect them as shown in Figure 6-3:

- renderers → shader → glsl_loader

- renderers → basic → textured_rectangle

- texture → loaders → png_tex_load

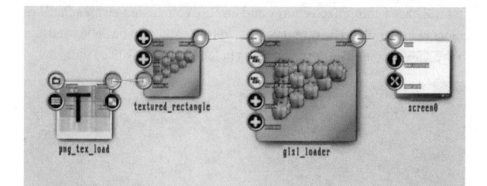

Figure 6-3. *Shader's basic state*

Note   You can find the sources at B-6.1_Color_Gradient in the TheArtOfAudioVisualization folder. You can find the shader code at /opt/thmad/share/thmad/ TheArtOfAudioVisualization-snippets/B-6.1_Color_ Gradient_Mapping.

Let module png_tex_load read a PNG, such as the architecture.png file included in the ThMAD distribution. Inside glsl_loader, write the following no-op vertex shader code:

```
void main(void) {
    gl_TexCoord[0] = gl_MultiTexCoord0;
    gl_Position = gl_ModelViewProjectionMatrix *

        gl_Vertex;
}
```

The fragment shader is more involved, as shown here:

```
#version 130
uniform sampler2D sampler;
uniform float blend; uniform float offset;

vec3 hsv2rgb(vec3 c){
    vec4 K = vec4(1.0, 2.0 / 3.0, 1.0 / 3.0, 3.0);
    vec3 p = abs(fract(c.xxx + K.xyz) * 6.0 - K.www);
    return c.z * mix(K.xxx,
                  clamp(p - K.xxx, 0.0, 1.0),
                  c.y);
}

void main() {


  const float[] grad = float[]( 0.58,0.98,0.9,0.58,0.98,0.88,
0.6,0.98,0.8,0.6,0.98,0.76,0.6,0.98,0.74,0.6,0.98,0.72,0.6,
0.98,0.78,0.58,0.98,0.8,0.58,0.98,0.82,0.58,0.98,0.84,0.58,
0.98, 0.86,0.58,0.94,0.88,0.58,0.94,0.9,0.58,0.92,0.9, 0.58,
0.9,0.9,0.58,0.88,0.9,0.58,0.88,0.92,0.58,0.86, 0.92,0.58,0.84,
0.92,0.58,0.82,0.92,0.58,0.82,0.94, 0.58,0.8,0.94,0.58,0.78,
0.94,0.58,0.8,0.92,0.58, 0.78,0.92,0.1,0.56,0.74,0.08,0.58,
0.74,0.08,0.56, 0.74,0.08,0.56,0.72,0.08,0.52,0.72,0.08,0.5,
0.72, 0.08,0.5,0.74,0.1,0.5,0.74,0.08,0.54,0.72,0.1,0.54, 0.74,
```

```
0.08,0.54,0.74,0.08,0.54,0.76,0.08,0.52,0.76, 0.08,0.52,0.74,
0.1,0.52,0.76,0.1,0.48,0.74,0.1,0.48, 0.76,0.1,0.46,0.76,0.08,
0.46,0.76,0.1,0.46,0.78, 0.08,0.44,0.76,0.56,0.48,0.96,0.58,
0.48,0.96,0.56, 0.48,0.94,0.56,0.46,0.94,0.56,0.42,0.96,0.58,
0.42, 0.98,0.58,0.42,0.96,0.56,0.42,0.98,0.56,0.4,0.98, 0.58,
0.38,0.98,0.56,0.36,0.98,0.56,0.32,0.98,0.56, 0.34,0.98,0.56,
0.34,0.96,0.56,0.38,0.98,0.56,0.4, 0.96,0.56,0.44,0.94,0.56,
0.44,0.96,0.56,0.46,0.96,0.56,0.5,0.94,0.56,0.5,0.96,0.56,0.52,
0.96,0.58,0.52,0.96,0.56,0.52,0.94,0.56,0.54,0.94,0.58,0.6,
0.96,0.58,0.62,0.96,0.58,0.64,0.96,0.58,0.66,0.96, 0.58,0.68,
0.96,0.58,0.68,0.94,0.58,0.72,0.94,0.58,

0.74,0.94,0.58,0.76,0.94,0.58,0.7,0.94,0.58,0.66, 0.94,0.08,
0.6,0.68,0.08,0.58,0.68,0.08,0.58,0.7, 0.08,0.6,0.72,0.08,
0.58,0.72,0.08,0.56,0.7,0.08, 0.54,0.7,0.1,0.5,0.76,0.08,0.5,
0.76,0.08,0.48,0.76,0.08,0.48,0.74,0.08,0.46,0.74,0.08,0.44,
0.74,0.08,0.42,0.74,0.08,0.42,0.76,0.08,0.4,0.76,0.08,0.38,
0.76,0.08,0.36,0.76
);

  vec4 tex = texture2D(sampler, gl_TexCoord[0].st);
  float gray = (tex.r + tex.g + tex.b)/3.0;
  gray = mod(gray+offset,1.0);

  int i = int(gray * 98.999);
  float ifract = gray * 98.999 - i;
  vec3 hsv1 = vec3( grad[i*3], grad[i*3+1],
                    grad[i*3+2] );
  vec3 hsv2 = vec3( grad[i*3+3], grad[i*3+4],
                    grad[i*3+5] );
  vec3 rgb = hsv2rgb(mix(hsv1,hsv2,ifract));

  gl_FragColor = mix(tex, vec4(rgb, tex.a), blend);
}
```

It acts as follows:

1. The first line tells which GLSL version is going to be used.

2. You declare the sampler2D uniform, which is mapped to the texture object.

3. Another uniform, blend, is used to mix the original texture and the texture with the color gradient applied.

4. The uniform offset is used to rotate the color gradient, presuming its end gets connected to its beginning in a circle.

5. The function hsv2rgb() is needed to convert from the HSV color space to the RGB color space.

6. Inside main(), grad is the array you got from the color gradient creation algorithm. It contains 100 HSV coordinates; thus, you have 300 elements.

Caution   Some graphics cards might complain about constant arrays not being allowed. In that case, try removing the const modifier.

---

7. Next you calculate the grayscale value of the texture pixel.

   For this calculation, the offset uniform is taken into account. gl_TexCoord[0].st fetches the vec2 coordinate vector from the associated texture (if you wrote .s, it would just be the first coordinate; with .t, it would be the second). Likewise, the .r, .g, and .b address the first, second, and third components of the tex vec4.

8. Then you write into i the index inside the gradient array, and into ifract the fraction to the next index.

9. The next lines calculate the adjacent HSV values.

10. The mix() function mixes the two HSV values you have so far using ifract and converts it into the RGB color space. You can see that mix() works for vectors just as for floats.

11. Finally, you mix (with mix()) the original texture and the texture with the gradient applied according to the blend uniform. If blend = 0.0, take only the original, and if blend = 1.0, take only the gradient color.

12. Write the result.

All that is left is to connect the two uniforms blend and offset to some controller modules, for example, input_visualization_listener. The output will look like, for example, Figure 6-4.

Figure 6-4. *Color gradient in action*

Of course, you could also feed a picture with the color gradient built by itself. Such an example could look like Figure 6-5.

Figure 6-5. *Picture using its own color gradient*

Shader Fractal

Fractals come in different forms and are the result of different algorithms. One algorithm you can use inside fragment shaders is a repeated mapping according to a couple of rules. For example, to determine the texture-based pixel value at some point P(x,y), you can write the following:

$$P'(x,y) = c_0(x,y) + c_1(f(x,y)) + c_2(f(f(x,y)))$$

$$+ c_3(f(f(f(x,y)))) + \ldots$$

Here, the $c_i()$ parts are color blend functions, and $f()$ is any suitable spatial transformation such as shifting, scaling, rotating, or a combination thereof. The function $f()$ could also be a mixture of $f_1()$, $f_2()$, ..., where each $f_i()$ gets applied with a certain probability w_i.

The vertex shader for such a visualization is the standard no-op shader, as shown here:

```
#version 130
void main(void)
{
    gl_TexCoord[0] = gl_MultiTexCoord0;
    gl_Position = gl_ModelViewProjectionMatrix
    *
gl_Vertex;
}
```

The fragment shader you are using reads as follows:

```
#version 130
uniform sampler2D sampler;
uniform float seed;
uniform float a;
uniform float b;
uniform float c;
```

```
float rand(float n){
    return fract(sin(n) * 43758.5453123);
}

void main() {
  mat3 f1 = mat3(a+1.609692,0.538246,0.070692,
    -0.531769,0.972514,0.515732,
    -1.645904,-0.627756,-1.527336);

  mat3 f2 = mat3(1.067971,b+0.122102,1.814637,
    -0.352147,0.404348,-1.815675,
    0.565247,c+0.024566,0.412591);

  vec3 v = vec3(gl_TexCoord[0].st +
       vec2(-0.1,-0.3), 0.0);
  vec3 col = vec3(0.0,0.0,0.0);
  float cc = 1.0;
  float r = seed;
  for(int i=0;i<20;i++) {

       v = r <= 0.5 ? f1 * v : f2 * v;
       vec4 tex = texture2D(sampler, v.xy);
       col += tex.rgb * cc;
       cc *= 0.9;
       r = rand(r);
  }
  col /= 3.0;

  gl_FragColor = vec4(col, 1.0);
}
```

It has two mappings, f_1 and f_2. The four uniforms seed, a, b, and c are for controlling from outside the shader. Specifically, the shader does the following:

1. It defines the two affine transform matrices f1 and f2, adding some dynamics by letting the uniforms a, b, and c change some of the parameters. The numbers used for the two matrices is the result of trial and error.

2. It defines a random generator. The algorithm here uses the fractional part of a heavily upscaled sine function. This is not a real random but comes pretty close.

3. It translates the basic texture coordinates vector by some constant vector and saves it in v.

4. It initializes a color accumulation buffer in c. The loop repeatedly applies one of f1 and f2 to v, determines the texture color at v, and adds it to c with decreasing intensity each loop iteration.

5. The division by 3.0 is to avoid over-saturation of the resulting fragment color.

6. It outputs the color accumulator c as the pixel color.

The state is the same as earlier in the chapter, with one additional module, texture → modifiers → tex_parameters, right between png_tex_load and textured_rectangle, with its wrap anchors set to clamp.

The output might look like Figure 6-6. The PNG used here for the texture is the feather.png file provided with the distribution.

Figure 6-6. *Shader fractal*

Note   You can find the source at B-6.2_Shader_Fractal in the TheArtOfAudioVisualization folder.

Timed Shader

Up to now you haven't used an explicit time inside the shaders. This is easy to accomplish. The system → time module provides for exactly what you need, and you can connect its normal / time output anchor to a uniform input anchor of the shader module.

Also, there is no real reason a shader *must* use the texture provided. You can also calculate pixel colors without using any texture pixel. The visualization you will be constructing here shows both, using explicit time and disregarding texture pixels.

The vertex shader code is again the no-op shader, as shown here:

```
#version 130
void main(void)
{
    gl_TexCoord[0] = gl_MultiTexCoord0;
    gl_Position = gl_ModelViewProjectionMatrix
    *
gl_Vertex;
}
```

For the fragment shader, write the following:

```
#version 130
uniform float
time;
uniform float colorful;
uniform float intensity;
uniform float phase;
uniform sampler2D
sampler;

float calcCol(int i, vec2 p, float l, float a) {
  float ang = 20*l*l*l;
  vec2 p2 = mat2(cos(ang),sin(ang),
                  -sin(ang),cos(ang)) * p;
  vec2 v = (p2+0.5) - p2 / l *
        (sin(a)+1.0) * abs(sin(l*9.0-a*1.0));
  return intensity /
        length( abs(mod(v,1.0) - vec2(0.5,0.0)) );
}
```

```
void main(){
  vec3 color;
  float z = time + phase;
  vec2 p = gl_TexCoord[0].st - 0.5;
  float l = length(p);

  for(int i=0;i<3;i++) {
    z += colorful; // shift RGB values
    color[i] = calcCol(i, p, l, z);
  }
  gl_FragColor = vec4(color/l,time);
}
```

Here is the explanation of the fragment shader:

1. There are three control parameters: time connected to the system time in seconds, colorful controlling the coloring (0 means no colors, 0.07 means normal, and 0.14 means high color), and phase serving as an offset to the time parameter. The uniform sampler gets connected to a texture; although you are not using its pixels, you still add it as an easy way to determine normalized window coordinates.

2. The calcCol() function calculates an R, G, or B color value depending on its parameters: the color index i (0 for red, 1 for green, 2 for blue), the pixel position vector p (out of [-1,-1]-[1,1] with (0.5,0.5) in the middle), its length, and a time-related variable a. With the angle ang, you introduce a swirl, which goes to the rotation-mat2 matrix; the rest consists of a couple of periodic functions defining the shape.

3. Inside main() you write to z the time plus offset,
to p the position vector referring to the center,
to l its length, and inside the loop you evaluate
the calcCol() function with a slightly shifted a
parameter for each color component. The color/l
expression intensifies the color near the center.

You can use the same state as in section "Color Gradient Algorithm"
above, with some control modules added for the shader uniforms. As a
texture, use any PNG you like (or use the blob) since you don't need its
pixel information.

Note   You can find the source at B-6.3_Timed_Shader in the
TheArtOfAudioVisualization folder.

Figure 6-7 shows the output.

Figure 6-7. *Timed shader*

Summary

In this chapter we invastigated a couple of more visualizations using shader constructs. We have seen how to apply a color gradient from inside a shader, and we introduced a shader fractal and a dynamic shader.

In the next chapter we will learn how to use ThMAD together withe the JACK audio sound server.

CHAPTER 7

ThMAD and the JACK Sound Server

ThMAD normally addresses the PulseAudio sound server. Using the JACK option switches, ThMAD can work with the JACK sound server as well, giving more options in a professional environment. This chapter describes how to use ThMAD with Jack.

Note that Ubuntu is not extraordinarily well suited to working with JACK because Ubuntu serves as a general-purpose desktop operating system and not as a dedicated audio- or video-authoring workstation. With an audio workstation, you would, for example, use a special real-time scheduling kernel with optimized low-latency audio handling. It is, however, possible to use JACK in Ubuntu, and I will show you how.

Using JACK for Sound Input

JACK is a sound server with extended routing capabilities and flexibility compared to ALSA or PulseAudio. However, using JACK while PulseAudio is running might lead to instabilities, so you might want to disable PulseAudio before you start using JACK.

Usually, disabling a server process is a matter of finding it and then stopping it, but for PulseAudio the story is a little bit more complicated. For improved stability, the developers of PulseAudio and the maintainers

© Peter Späth 2018
P. Späth, *Advanced Audio Visualization Using ThMAD*,
https://doi.org/10.1007/978-1-4842-3504-1_7

of Ubuntu added an autospawn functionality, meaning the server process is being observed, and once it disappears from the process list, it gets restarted automatically once a client process tries to use PulseAudio.

Fortunately, there is a better way of addressing this issue without the need to disable PulseAudio. A program named pasuspender can temporarily suspend the PulseAudio server from accessing devices so other processes will be able to access them without PulseAudio interfering. You will be using that suspender in this chapter.

One more step you need to do the first time you set up JACK for ThMAD is to add your Ubuntu Linux user account to the audio group. First, check whether you are already a member of that group by entering groups in a terminal, and if the list that then appears contains *audio*, you are already done. If not, change to root via sudo su and then enter usermod -a -G audio [USER], where [USER] is the name of your account (if the group doesn't exist, enter groupadd audio first). Log out and in again for the changes to take effect.

Next, you need to install a couple of tools you will be using. Switch to the root user and enter the commands shown in Table 7-1. Entering the install commands if you already have the package in question does not hurt, so you can try this without first checking. The format is as follows:

apt-get install [PACKAGE]

Table 7-1. *Tools*

Tool	Description
alsa-utils	Some utilities for the underlying sound architecture
jackd	The JACK sound server
jack-tools	Some tools for JACK
qjackctl	GUI for the JACK server

Now start the JACK server control GUI by entering a terminal, as shown here:

```
pasuspender qjackctl
```

This will start the administration tool while PulseAudio gets temporarily suspended. The main GUI will appear, as shown in Figure 7-1.

Figure 7-1. *Qjackctl main GUI*

Caution   Once installed, the Qjackctl program can also be started using the Ubuntu starter/launcher. However, this will not start it with PulseAudio suspended, so you *must* use the terminal and enter the command shown earlier.

You might be able to start the JACK server by just clicking the Start button, but you usually have to first use the Setup button and change a few settings there, as shown in Figure 7-2. You should not select Realtime unless you know what you do. Also, a sample rate of 48000 should do, but in case you run into trouble, you could also try 44100. Clicking the Advanced tab will show the dialog in Figure 7-3. Select No Memory Lock. The fields Output Device and Input Device are crucial because they point to the sound card to use. You could try one of the entries from the drop-down list, but to get a better idea of what to enter here, enter aplay -l in a terminal to get a list of sound devices on your system.

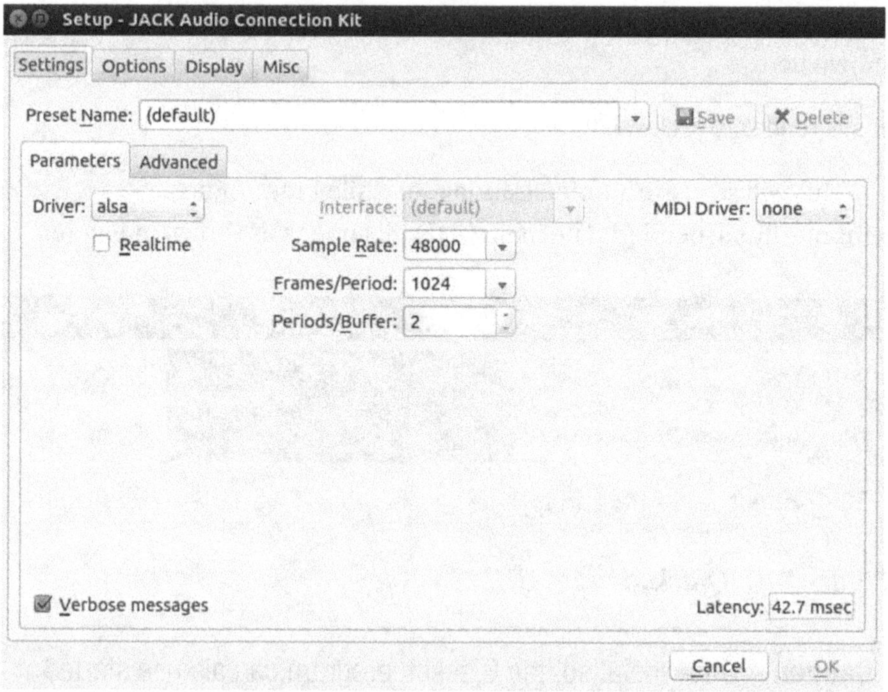

Figure 7-2. *Setting up Qjackctl, Parameters tab*

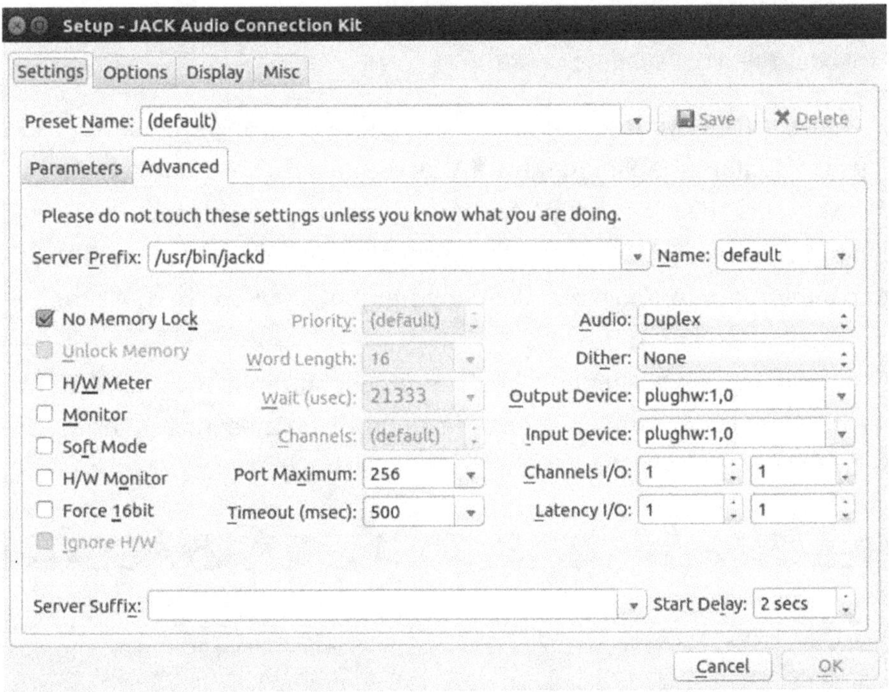

Figure 7-3. *Setting up Qjackctl, Advanced tab*

This shows a list of devices like you will see:

```
**** List of PLAYBACK Hardware Devices ****
card 0: HDMI [HDA Intel HDMI], device 3: HDMI 0 [HDMI 0]
  Subdevices: 1/1
  Subdevice #0: subdevice #0
card 0: HDMI [HDA Intel HDMI], device 7: HDMI 1 [HDMI 1]
  Subdevices: 1/1
  Subdevice #0: subdevice #0
card 0: HDMI [HDA Intel HDMI], device 8: HDMI 2 [HDMI 2]
  Subdevices: 1/1
  Subdevice #0: subdevice #0
card 1: PCH [HDA Intel PCH], device 0: ALC269VB Analog
[ALC269VB Analog]
```

```
Subdevices: 1/1
Subdevice #0: subdevice #0
```

Once you find your sound card, say a card called X and a subdevice called Y, you can enter **hw:X,Y** or **plughw:X,Y** for the fields in the Qjackctl GUI.

Now click the Start button, and if no error message appears, you can probe the server. To do so, in a terminal enter `jack_metro -b 60` to start a metronome. You won't hear anything yet, however, because the connections first must be set.

To connect JACK clients, click the Connect button in the main GUI to see the Connections dialog; then drag and drop connections as shown in Figure 7-4.

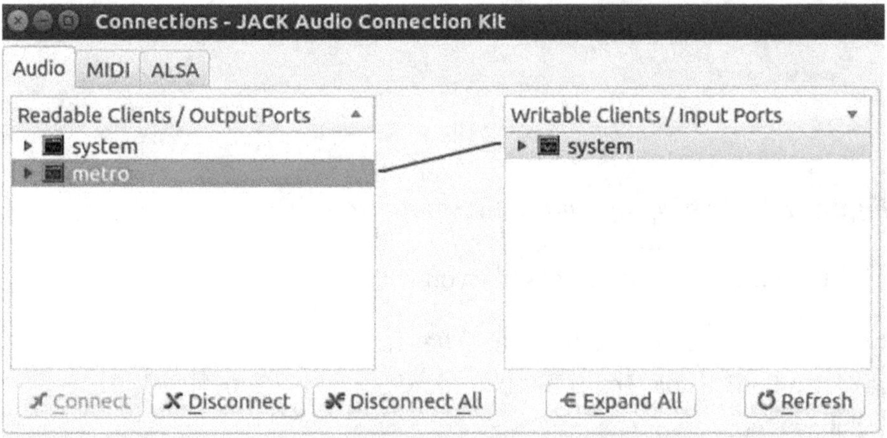

Figure 7-4. *Connections in Qjackctl*

You should now hear the metronome beeping. In the next section, you will connect ThMAD to the running server.

ThMAD and JACK Together

To work with JACK, ThMAD needs a stereo input device right from the beginning. So, you cannot, for example, use the `jack_metro` command from the previous section because that one produces only mono.

You can provide a "thru" device that can serve as a stereo input device to ThMAD and later connect any sound input to that "thru" device since it just hands sound data unchanged through. To start a "thru" device, open a terminal and enter the following:

```
jack_thru
```

The Qjackctl GUI will show a new readable and a new writable client in the Connections dialog after you click the Connect button; see Figure 7-5.

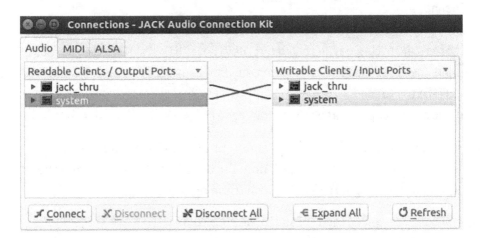

Figure 7-5. *Qjackctl with the jack_thru client*

Remove the connection from `system` as a readable client to `jack_thru` as a writable client because you want to later use a different input to `jack_thru`.

You can now tell ThMAD to connect to JACK by using the following startup command:

```
/opt/thmad/thmad_artiste -sound_type_jack
```

Look at the output ThMAD produces in the terminal. If JACK is running and the jack_thru client is registered, the output should contain something like this:

```
rtaudio_record.h
audioprobe() Audio Type =
Jack Client Available APIs:
  Jack Client
  Linux ALSA

  Linux PulseAudio Current API: Jack Client [...]
Found 2 device(s) ... [...]
DEVICE NUMBER = 1
Device Name = jack_thru Probe Status = Successful Output
Channels = 2
Input Channels = 2 Duplex Channels = 2
This is NOT the default output device. This is NOT the default
input device. Natively supported data formats:
  32-bit float
Supported sample rates = 48000 [...]
```

Once you have identified the sound producer device you need (here device number 1 for jack_thru), note the DEVICE NUMBER value and the sample rate and restart ThMAD with these options:

```
-sound_type_jack -snd_rtaudio_device=1 \
-snd_sample_rate=48000
```

If the device happens to be the default device, you can omit the snd_rtaudio_device switch, and if the sample rate you want to use is 44,100, you can also omit the snd_sample_rate switch. In the example, you can see 44,100 is not allowed, but 48,000 is, so the sample rate of 48,000 has to be specified there.

ThMAD is now running and connecting to the jack_thru device, and you can connect any JACK sound producer to jack_thru, which then just forwards the audio data to ThMAD. The Connections dialog from Qjackctl will show something like Figure 7-6.

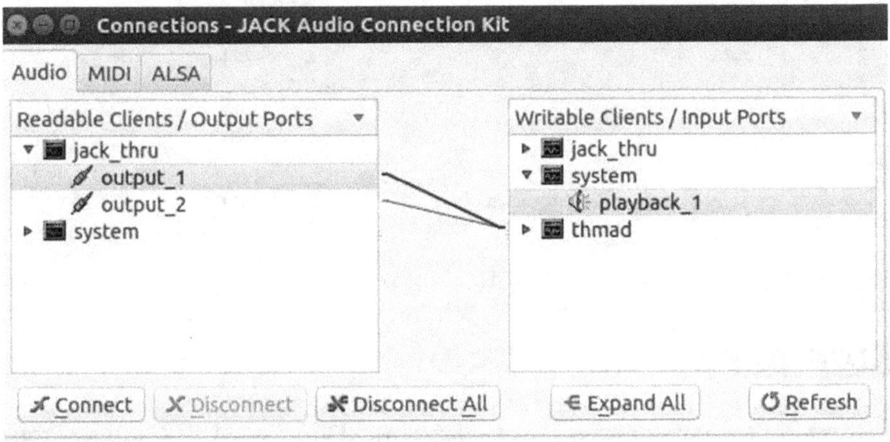

Figure 7-6. *ThMAD as a JACK client*

If you want to test it with the simple metronome client, start that again via jack_metro -b 60 and then draw the connection from metro to jack_thru in the Connections dialog, as shown in Figure 7-7. Load a visualization into ThMAD, and you should then see ThMAD react to the metronome beeps.

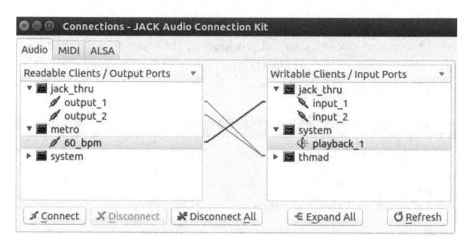

Figure 7-7. *ThMAD and the JACK metronome*

Summary

In this chapter, you learned how to use ThMAD inside a JACK audio server system. You saw how to set up appropriate parameters and how to use a couple of tools to get ThMAD talking with JACK.

In the next chapter, you will learn how to let ThMAD directly use the ALSA API.

CHAPTER 8

TMAD and ALSA

If you don't want to use PulseAudio as a sound server and instead want
to connect to the underlying Linux sound architecture ALSA directly,
you need to take a couple of things into account. First, using ALSA while
PulseAudio is running might cause problems. But even more important,
you need to find out the correct parameters for using ALSA, which is not
always easy.

Disabling PulseAudio

To make sure PulseAudio is not running, in a terminal (press Ctrl+Alt+T to
get one) enter the following:

```
pulseaudio --kill
```

This stops PulseAudio. Then enter the following to check whether
PulseAudio is still running:

```
ps auxw | grep pulseaudio | grep -v grep
```

If this command does *not* produce a line like the following, it means
you are good and PulseAudio is no longer running:

```
user 8335 0.0 0.0 639808 11964 ? S<l 11:54
0:00 pulseaudio ...
```

© Peter Späth 2018
P. Späth, *Advanced Audio Visualization Using ThMAD*,
https://doi.org/10.1007/978-1-4842-3504-1_8

If instead PulseAudio still shows up, you probably have the autospawning function of PulseAudio enabled. To disable it, in the file /etc/pulse/client.conf, look for a line like autospawn = yes and make sure it has a semicolon at the beginning. Underneath it, add the line autospawn = no *without* the semicolon. Note that you need to be logged in as root to make changes in this file (enter sudo su first).

Save the file and restart your computer. Now running the check with the ps ... command shown earlier should produce no output. Note that with autospawning off, you must manually control PulseAudio. If you need it, enter pulseaudio --start to start it and pulseaudio --kill to stop it.

Starting ThMAD with ALSA

Explaining how to set up ALSA is beyond the scope of this book. The rest of this chapter assumes you have set up ALSA correctly to allow for capturing input.

For further configuring ThMAD, install the alsa-utils package, which contains some helpful tools. The following needs to be entered as root:

```
apt-get install alsa-utils
```

Now exit root by pressing Ctrl+D and then enter aplay -l. The output will look like this:

```
...
card 2: PCH [HDA Intel PCH], device 0: ALC269VB Analog
[ALC269VB Analog]
  Subdevices: 1/1
  Subdevice #0: subdevice #0
...
```

Try to identify a suitable input device in the output. The line you see here, for example, points to my built-in microphone. Deduce a hardware ID from that; from the sample, take 2 from card 2 and 0 from Subdevice #0, and enter hw:2,0. Try to see whether this device receives input. In a terminal, enter the following and produce input (speak into the microphone or play music if this is another capturing device):

```
arecord -D hw:2,0 -vv -f dat /dev/null
```

If the output reacts to sound input, this is the device you need to be looking at.

If you see the arecord program working, as shown in Figure 8-1, but it doesn't react to input, the percentage stays at 0 percent.

Figure 8-1. *arecord sound input*

Even with input, maybe the sound capture is not enabled. Use alsamixer in a terminal to possibly unmute and start a capturing device.

```
alsamixer -c2
```

Here, 2 is the device number. With alsamixer running, use F4 to switch to capturing devices, use the arrow keys to select the capture setting, use m to possibly unmute, and use the spacebar to start the capturing. See Figure 8-2.

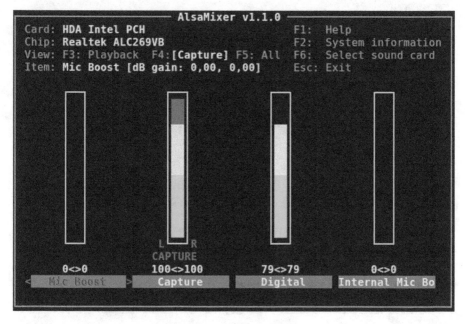

Figure 8-2. *The alsamixer tool*

Try the following again, and now you should see it react to sound:

```
arecord -D hw:2,0 -vv -f dat /dev/null
```

If it shows 99 percent all the time, you must change the gain. Still in alsamixer with the Capture control selected, use the up arrow or down arrow to change the capture gain. The correct output of the `arecord` program looks like Figure 8-3.

Figure 8-3. *arecord sound input working*

Now you need to tell ThMAD to use that ALSA device. Make sure the `arecord` program from earlier has been stopped (ALSA doesn't like concurrent device usage), and also make sure the last state you saved

contains the input_visualization_listener module. Now start ThMAD
Artiste in a terminal, but with the following at the end:

```
-sound_type_alsa
```

After it starts, quit the program. The terminal now contains diagnostic
output. More precisely, it will list all the ALSA devices it sees. In the listing,
identify your device. For example, for me it contains the following:

```
[...]
DEVICE NUMBER = 5
Device Name = hw:HDA Intel PCH,0 Probe Status =
Successful  Output Channels = 2
Input Channels = 2 Duplex Channels = 2
This is NOT the default output device. This is NOT the default
input device. Natively supported data formats:
  16-bit int

  32-bit int
Supported sample rates = 44100 48000 96000
192000 [...]
```

By comparing the device name hw:HDA Intel PCH,0 with the output
from aplay -l shown earlier, card 2: PCH [HDA Intel PCH], ..., you
can see that the fifth device (DEVICE NUMBER = 5) is the one that ThMAD
needs to be told to use. To do this, start ThMAD in a terminal and add the
following at the end:

```
-sound_type_alsa -snd_rtaudio_device=5
```

If you omit the second switch with the device ID, the default device
gets used instead.

If the device in the listing that ThMAD produces does not contain a sample rate of 44,100, you need to tell ThMAD to use a different sample rate. You can do this by adding another option while starting the program. For example, enter the following options:

```
-sound_type_alsa -snd_rtaudio_device=5
-snd_sample_rate=48000
```

This tells ThMAD to use ALSA, device number 5, and a sample rate of 48,000 instead.

Summary

In this chapter, you learned how to let ThMAD use ALSA directly instead of PulseAudio.

In the next chapter, you will learn how to control ThMAD from the outside using scripts or other programs.

CHAPTER 9

Controlling ThMAD from the Outside

Whenever you perform any action inside ThMAD Artiste, an internal messaging framework sends appropriate messages to a rendering engine working inside ThMAD. While by design the engine will not run without Artiste or Player showing the graphics, the state creation itself can be outsourced to an external script or program.

In this chapter, you will learn how ThMAD can be configured to receive messages from the outside, and you will see how a client can be used to create such messages.

ThMAD and Its Server Socket

You know that both ThMAD Artiste and ThMAD Player can be controlled from the outside by certain modules such as the sound listener, and of course you can use the Artiste GUI to manually set the value of any input anchor to control the sketch. Wouldn't it be nice if you had a general-purpose interface for setting anchors from the outside? You could have another software or script running that then feeds that interface. By doing this, you would get an enormous boost in the possibilities for interesting visualizations. The price you have to pay for this is that you need to learn another programming language or find a way to let some other software talk to ThMAD in the language it understands.

© Peter Späth 2018
P. Späth, *Advanced Audio Visualization Using ThMAD*,
https://doi.org/10.1007/978-1-4842-3504-1_9

The good news is ThMAD provides such an interface in the form of a *server socket*. Think of a server socket as some low-level communication endpoint with a standardized way to connect to it. Web browsers, for example, internally connect to server sockets running on some remote server machines. The analogy goes further: to connect to a web server, the browser needs its Internet address and a port, and everyone knows how to specify this via a URL like this: `http://some.funky.server.com:80`.

Here, `:80` denotes the port, and since 80 is the default, it can be quite often omitted in the address specification. A client trying to talk to ThMAD will also need to know the address of ThMAD's server socket. That is done using the following format while starting Artiste or Player from a terminal:

```
-port NUMBER
```

Here's an example:

```
/opt/thmad/thmad_artiste -port 30533
```

There are three basic things you have to know before you start ThMAD with a server socket running. First, any port below 1024 is reserved for the operating system and should not be used for applications. Why then are web servers running on port 80? The full story is that HTTP was given such importance from the very beginning that it can run under any of the reserved ports from 1 to 1023. It is not forbidden that an application runs on a port under 967, for example; it is just not allowed for nonroot processes, and since you run ThMAD as a nonroot process, it must use one of the ports starting at 1024. Second, the maximum port number that can be used is $2^{16}-1 = 65535$. Third, and maybe trickiest, no other applications can concurrently use the same port. Usually high numbers are a good bet, like 30174. If the port is unavailable because it's being used by some other application, ThMAD will tell you via some startup error message like this:

```
Exception was caught: Could not bind to port.
```

Then you can try another one. Note that under such circumstances ThMAD will still start up despite the port clash, but the socket specified under -port will not start, and you hence cannot talk to ThMAD from the outside.

You can also make a more educated guess and first create a listing of used ports by entering the following:

```
netstat -lntu
```

The numbers after the colon (:) are the port numbers.

ThMAD Socket Clients

For the client application, which is the application that connects to ThMAD from the outside, you can use any sufficiently elaborated platform like Java or a scripting language like Groovy or Python. Talking to server sockets is such a basic and standardized process that you will be able to find lots more clients to use for this purpose. As an example, as Groovy is one of my favorites, a sample code snippet for talking to ThMAD from any server connected via the Internet in Groovy looks like this:

```groovy
SERVER_ADDR = 'localhost'
SERVER_PORT = 31567
def s = new Socket(SERVER_ADDR, SERVER_PORT)
def res = ''
s.withStreams { input, output ->
        output << "<COMMAND>"
        def reader = input.newReader()
        def buffer = reader.readLine()
        res = buffer.trim()
}
s.close()
```

Here, SERVER_ADDR is the Internet address of the machine where ThMAD is running, SERVER_PORT is the port, and <COMMAND> is the command to be executed. The output of the command is stored in the variable res.

You can also use a terminal. After you open one (e.g., by pressing Ctrl+Alt+T), you can fire a command like this:

```
echo "<SOME COMMAND>" | nc localhost 32111
```

Here, localhost 32111 specifies that ThMAD is running on the same machine and listening at port 32111. Replace localhost with any Internet address and use a port number different from 32111 if this is not the case.

Messages

Table 9-1 describes the language idioms you can use to talk to ThMAD with a running server socket.

Table 9-1. *Messages*

Message	Description
show state_name	Returns the complete file system path of the currently active state or visual.
show state	Returns the complete state as a list of commands, like in a state file.
show meta_information	Returns the meta information associated with a state. The meta information is what you added in the optional fields when saving a state in ThMAD Artiste.
get param <MOD> <PAR>	Returns the current value of the input anchor PAR of module MOD. The module name is the same as you see in ThMAD Artiste right underneath the module symbol, or it's the third column of a component_ create line inside a state file. Currently available for INT, FLOAT, FLOAT3, FLOAT4, and QUATERNION type anchors. For FLOAT3, FLOAT4, and QUATERNION, the output will be comma-separated.
set param <MOD> <PAR> <VAL>	Sets the input anchor PAR of module MOD to the value given. The module name is the same as you see in ThMAD Artiste right underneath the module symbol, or it's the third column of a component_ create line inside a state file. Currently available for INT, FLOAT, FLOAT3, FLOAT4, and QUATERNION type anchors. For FLOAT3, FLOAT4, and QUATERNION, the input VAL will have to be a comma-separated list.
cmd ...	Sends a command string to the internal message queue.

For example, to set the gamma_correction input anchor of screen0 to 1.3, you'd write the following:

```
echo "set_param screen0 gamma_correction 1.3" | nc localhost 32111
```

The engine will immediately set the parameter as specified.

Raw Commands

In addition to controlling ThMAD by setting parameters, it is possible to simulate all the other ThMAD Artiste GUI activities using the server socket. Any command that starts with "cmd" (with the trailing space) will be stripped of the "cmd" part, and the rest will be going unchanged into the same internal command processing queue Artiste is using for its work. Even better, since state files contain nothing but lists of such commands, you can look at any sample state and use the commands found there to feed the server socket. Just add "cmd" at the beginning of each command.

For example, creating a module via a socket can be as easy as executing this in a terminal:

```
echo "cmd component_create renderers;basic;colored_rectangle
        colored_rectangle7 -0.150612 0.014989" |
nc localhost 32111
```

Thus, you can construct complete states using nothing but Artiste's server socket! This is not possible with Player, though, since it always handles complete states.

The internal command structure can be deduced by looking at state files with a text editor. The only exception is the following, which is used to delete modules and will naturally not show up in the state files:

```
component_delete <alias>
```

Table 9-2 provides a detailed description of the command parameter type representations.

Table 9-2. *Raw Command Parameter Types*

Type	Representation
float	The usual float representation, for example: 123.56 or -1743.0.
float3, float4	A comma-separated list of floats, for example: 0.1,2.0,-5.4. Do not add spaces anywhere!
quaternion	A comma-separated list of four floats, for example: i,j,k,w. Do not add spaces anywhere!
enumeration	An int specifying the position inside the enumeration: 0 is the first, 1 the second, and so on.
resource	A Base64-encoded string starting from, and including, the resources folder.
shader code	A Base64-encoded string containing the shader code.
sequence	A pipe-separated list of sequence points. Each point looks like x-len;type;y-val, where x-len is the x-distance to the next point, type is the interpolation type (0.0 = no, 1.0 = linear, 2.0 = cosine, 4.0 = Bezier), and y-val is the Base64-encoded y-value. Possible values are a single float for no, linear, or cosine interpolation; a semicolon-delimited list of five floats for Bezier-interpolation; and the following: – y-value of the point – x-distance of the first control point – y-distance of the first control point – x-distance of the second control point (to the next point!) – y-distance of the second control point (to the next point!) The last point has always x-len=1.0 set.

Summary

In this chapter a way to control and even build visualizations from outside ThMAD using a server socket got described.

In the next chapter we will describe advanced configuration issues which will help you to configure ThMAD according to your needs.

CHAPTER 10

Configuration

ThMAD Artiste has a couple of configuration entries. This chapter will describe how to access and change them.

Accessing the Configuration

You can access the configuration directly from ThMAD Artiste using the main pop-up menu (Figure 10-1).

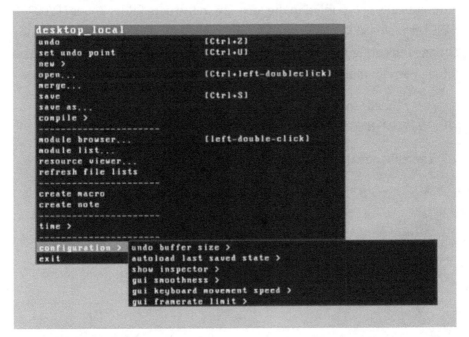

Figure 10-1. *Configuration menu*

© Peter Späth 2018
P. Späth, *Advanced Audio Visualization Using ThMAD*,
https://doi.org/10.1007/978-1-4842-3504-1_10

If you like, you can also change it by editing the configuration file under this path:

```
/home/[YOUR_NAME]/thmad/ [VERSION]/data/thmad.conf
```

or this path:

```
/home/[YOUR_NAME]/.local/share/thmad/ [VERSION]/data/thmad.conf
```

The first is a symbolic link to the second, so they point to the same file.

If you search for that file and can't find it, there is a simple explanation. Upon startup, ThMAD first looks for that file, and if it can't find it, it will look for a file called thmad.conf inside its installation, shown here:

```
/opt/thmad/share/thmad/thmad.conf
```

It will work with this file unless you changed the configuration from inside ThMAD Artiste. If you changed it, the file from the installation automatically gets copied to the location in the data folder /home/[YOUR_NAME]/thmad/[VERSION]/data and then changed there. So, if you want to change the configuration manually and you cannot find the file in the data folder, you can make that copy by yourself.

```
cp /opt/thmad/share/thmad/thmad.conf
/home/[YOUR_NAME]/thmad/[VERSION]/data
```

Then change the configuration using a text editor.

Configuration Entries

Table 10-1 describes all the configuration entries.

Table 10-1. *Configuration File Entries*

Key in the File	Pop-up Menu Item	Description
assistant_size	None. But you can toggle through a set of different sizes using the Tab key.	The size of the ThMAD Artiste Assistant.
autoload_last_ saved_state	autoload last saved state → no / yes	If 1, upon Artiste startup, automatically loads the last saved state. If 0, loads the _default state.
global_framerat e_limit	gui_framerate limit → none / * fps	Sets an upper limit to the frame rate. If -1, no limit applies. Otherwise, the redrawing frequency for both the GUI and the output will be limited to the given number.

global_ interpolation_ speed	gui_smoothness → *	Sets how movement and sizing operations on the GUI will be interpolated. The following values can be used:
		  − 1000.0: No interpolation (or None in the pop-up menu)
		  − 2.0: Quick
		  − 1.0: Normal
		  − 0.5: Slow

(continued)

Table 10-1. (*continued*)

Key in the File	Pop-up Menu Item	Description
global_key_ speed	gui keyboard movement speed → *	Controls the panning and zooming speed if the keyboard gets used (S, D, F, W, E, R or cursor/PgUp/ PgDown keys): – 3: Normal – 2: Slow – 1: Very slow
global_show_ inspector	show inspector → *	Sets whether to show the inspector (error: "Reference source not found"); 1 shows it, and 0 does not.
last_saved_ state	None	Sets the Base64-encoded path to the last saved state. Applies only if configuration item autoload_ last_saved_state is set to 1.
skin	None	Sets which skin (a GUI's appearance, mainly colors) to use. Currently ThMAD uses only one default skin named thmad_ plain, so you should not change this value.
undo_ buffer_size	undo buffer size → *	Sets how many undo operations are possible.

Summary

In this chapter we investigated a few configuration tweaking methods beyond just using the GUI.

Index

A

ALSA
  PulseAudio, 205–206
  ThMAD, 206–207, 209
Ambient light, 154–156
Artiste operation, ThMAD
  Artiste files, 4–5
  fullwindow mode, 2–3
  options, 2–3
  performance mode, 4
  stopping method, 4

B

Bitmap texture distortion
  blob parameters, 77
  coordinates, 76
  description, 75
  3D pipeline, 77
  module mesh_tex_bitmap_
      distort, 77
  mesh tex bitmap distort
      parameters, 78
  output, 78
  RED and GREEN values, 76

C

calcCol() function, 192
Candle-like fire, 128
Clear color, 154
Color gradient
  action, 185
  algorithm, 172–178
  calculated, 179
  extraction, 172
  mapping, 171
  pictures using, 186
Cube mesh, 21–23

D

Diffuse light, 154, 156–157

E

Emissive light, 159–161
Engine states
  ENGINE_LOADING, 39
  ENGINE_PLAYING, 39, 40
  ENGINE_STOPPED, 39, 40
  ENGINE_REWIND, 39

© Peter Späth 2018
P. Späth, *Advanced Audio Visualization Using ThMAD*,
https://doi.org/10.1007/978-1-4842-3504-1

V, W, X, Y, Z

Get the eBook for only $5!

Why limit yourself?

With most of our titles available in both PDF and ePUB format, you can access your content wherever and however you wish—on your PC, phone, tablet, or reader.

Since you've purchased this print book, we are happy to offer you the eBook for just $5.

To learn more, go to http://www.apress.com/companion or contact support@apress.com.

Apress®

CPSIA information can be obtained
at www.ICGtesting.com
Printed in the USA
LVHW01s1753010418
571883LV00005B/42/P

9 781484 235034